Spatial Variabilities of Soils and Landforms

Spatial Variabilities of Soils and Landforms

Proceedings of an international symposium sponsored by Division S-5 of the Soil Science Society of America and the International Society of Soil Science in Las Vegas, Nevada, 17 Oct. 1989.

Editors
M. J. Mausbach and L. P. Wilding

Organizing Committee
M. J. Mausbach and L. P. Wilding

Editorial Committee
M. J. Mausbach, *chair*, L. P. Wilding, E. M. Rutledge, L. F. Ratliff

Editor-in-Chief SSSA
R. J. Luxmoore

Associate Editor-at-Large
J. M. Bartels

SSSA Special Publication Number 28

**Soil Science Society of America, Inc.
Madison, Wisconsin, USA
1991**

Cover art: soil survey map provided by Dr. M.J. Mausbach

Copyright © 1991 by the Soil Science Society of America, Inc.

ALL RIGHTS RESERVED UNDER THE U.S. COPYRIGHT LAW OF 1978 (P.L. 94-533)

Any and all uses beyond the limitations of the "fair use" provision of the law require written permission from the publisher(s) and/or the author(s); not applicable to contributions prepared by officers or employees of the U.S. Government as part of their official duties.

Second Printing 1992

Soil Science Society of America, Inc.
677 South Segoe Road, Madison, WI 53711, USA

Library of Congress Cataloging-in-Publication Data

Spatial variabilities of soils and landforms / editors, M.J. Mausbach and L.P. Wilding.
    p.   cm. — (SSSA special publication : no. 28)
  "Proceedings of an international symposium sponsored by Division S-5 of the Soil Science Society of America and the International Society of Soil Science in Las Vegas, Nevada, 17 October 1989."
  Includes bibliographical references.
  ISBN 0-89118-798-7
  1. Soil surveys—Congresses. 2. Soil mapping—Congresses. 3. Soil geography—Congresses. 4. Landforms—Congresses. I. Mausbach, Maurice II. Wilding, Larry P. III. Soil Science Society of America. Division S-5. IV. International Society of Soil Science. V. Series.
S592.14.S66   1991
631.4'7—dc20   91-26871
                                                                         CIP

Printed in the United States of America

CONTENTS

Foreword vii
Preface ix
Contributors xi
Conversion Factors for SI and non-SI Units xiii

1 The Need to Quantify Spatial Variability
   R. W. Arnold and L. P. Wilding 1

2 Predicting Variability of Soils from Landscape Models
   G. F. Hall and C. G. Olson 9

3 One Perspective on Spatial Variability in Geologic Mapping
   H. W. Markewich and S. C. Cooper 25

4 Scientific Methodology of the National Cooperative Soil Survey
   Stephen L. Hartung, Steven A. Scheinost,
   and Robert J. Ahrens 39

5 Statistical Procedures for Specific Objectives
   Dan R. Upchurch and William J. Edmonds 49

6 A Comparison of Statistical Methods for Evaluating Map Unit Composition
   S. C. Brubaker and C. T. Hallmark 73

7 Sampling Designs for Quantifying Map Unit Composition
   P. A. Burrough 89

8 Presentation of Statistical Data on Map Units to the User
   R. B. Brown and J. H. Huddleston 127

9 Soil Mapping Concepts for Environmental Assessment
   Duane A. Lammers and Mark G. Johnson 149

10 Minimum Data Sets for Use of Soil Survey Information in Soil Interpretive Models
   R. J. Wagenet, J. Bouma, and R. B. Grossman 161

11 Quantifying Map Unit Composition for Quality Control in Soil Survey
   L. C. Nordt, John S. Jacob, and L. P. Wilding 183

12 Using Systematic Sampling to Study Regional Variation of a
    Soil Map Unit
        G. W. Schellentrager and J A. Doolittle 199

13 Confidence Intervals for Soil Properties within Map Units
        Fred J. Young, J. M. Maatta, and R. David Hammer 213

14 Spatial Variability of Organic Matter Content in Selected
    Massachusetts Map Units
        M. Mahinakbarzadeh, S. Simkins, and P. L. M. Veneman   231

15 Geographic Information Systems for Soil Survey and
    Land-Use Planning
        R. David Hammer, Joseph H. Astroth, Jr.,
        G. S. Henderson, and Fred J. Young 243

FOREWORD

Spatial variability is an often neglected consideration in field studies of natural resources. It is important that the full power of statistics and data processing technology be applied to this area that we have known and described in a qualitative fashion in the past. The authors of this volume have made a significant contribution to the understanding and development of techniques for describing variability in soils and landforms. Advances in these techniques have been great in recent years. This volume brings together much of this new knowledge and demonstrates its application in natural settings. The Soil Science Society of America is pleased to make this work available to all disciplines that utilize soil information.

WILLIAM W. MCFEE, *president-elect*
Soil Science Society of America

PREFACE

As technology for displaying and integrating soil geographical data into natural resource assessment expands, the demands for more precise information in support of soil resource inventories also increase. The need for such knowledge becomes even more acute as we enter the era of computers and information systems. Hence, users of soil surveys are demanding quantitative information on the composition and variability of soils in map unit delineations. This is true of surveys produced by the National Cooperative Soil Survey in the USA and elsewhere in the world. In short, soil surveys are under interrogation; the challenge is for soil scientists to provide sufficient verification to meet increased information demands.

The objective of this symposium was to establish a forum for creative proposals in developing guidelines and procedures for quantitative description of map unit composition. Speakers were invited to illustrate relationships of landforms, landscapes, and stratigraphy to soil spatial variability, elucidate the scientific methodology as applied to soil surveys, compare and contrast the utility of classical and regionalized theory and statistical approaches in defining soil spatial variability, and present alternative strategies for transferring spatial variability knowledge to natural resource user clientele. Additionally, volunteer papers were included that expanded on the above topical thrusts and/or represented specific case studies for quantifying soil variability in map units.

Papers included in this document capture the state-of-the-art in quantifying soil/landform features. Challenges are presented for the scientific community to develop quality assurance control procedures for soil surveys and for leaders of the National Cooperative Soil Survey Program to implement these procedures into soil survey standards and procedures.

M.J. MAUSBACH, *symposium cochair*
USDA-SCS, Soil Survey Division,
Washington, District of Columbia

L.P. WILDING, *symposium cochair,*
Department of Soil and Crop Sciences,
Texas A&M University,
College Station, Texas

CONTRIBUTORS

Robert J. Ahrens	Soil Scientist, USDA-SCS, Lincoln, Nebraska
R. W. Arnold	Director, Soil Survey Division, USDA-SCS, Washington, District of Columbia
Joseph H. Astroth, Jr.	Assistant Professor of Geography, Department of Geography, University of Missouri–Columbia, Columbia, Missouri
J. Bouma	Professor of Soil Science, Department of Soil Science, Agricultural University, Wageningen, the Netherlands
R. B. Brown	Professor of Soil Science, Soil Science Department, University of Florida, Gainesville, Florida
S. C. Brubaker	Research Associate, USDA-ARS, Lincoln, Nebraska
P. A. Burrough	Professor of Physical Geography, Institute of Geographical Science, University of Utrecht, Utrecht, the Netherlands
S. C. Cooper	Senior Project Geologist, Blasland, Bouck & Lee Engineers, Syosset, New York
J. A. Doolittle	Soil Specialist, USDA-SCS, Chester, Pennsylvania
William J. Edmonds	Associate Professor, Virginia Polytechnic Institute and State University, Blacksburg, Virginia
R. B. Grossman	Soil Scientist, SCS, National Soil Survey Center, Lincoln, Nebraska
G. F. Hall	Professor, Department of Agronomy, Ohio State University, Columbus, Ohio
C. T. Hallmark	Associate Professor of Pedology, Texas Agricultural Experiment Station, Department of Soil and Crop Sciences, College Station, Texas
R. David Hammer	Assistant Professor of Soil Science, School of Natural Resources, University of Missouri, Columbia, Missouri
Stephen L. Hartung	Research Soil Scientist, Conservation and Survey Division, University of Nebraska-Lincoln, Lincoln, Nebraska
G. S. Henderson	Professor of Forest Soils, School of Natural Resources, Univesity of Missouri, Columbia, Missouri
J. H. Huddleston	Professor of Soil Science, Department of Soil Science, Oregon State University, Corvallis, Oregon
John S. Jacob	Research Associate, Department of Soil and Crop Science, Texas A&M University, College Station, Texas
Mark G. Johnson	METI, Inc., USEPA Environmental Research Laboratory, Corvallis, Oregon
Duane A. Lammers	Soil Scientist, USDA-FS, USEPA Environmental Research Laboratory, Corvallis, Oregon

J. M. Maatta	Assistant Professor of Statistics, Department of Statistics, University of Missouri, Columbia, Missouri
M. Mahinakbarzadeh	Research Associate, Department of Plant and Soil Sciences, University of Massachusetts, Amherst, Massachusetts
H. W. Markewich	Geologist, U.S. Geological Survey, Doraville, Georgia
M.J. Mausbach	USDA-SCS, Soil Survey Division, Washington, District of Columbia
L. C. Nordt	Lecturer, Department of Soil and Crop Sciences, Texas A&M University, College Station, Texas
C. G. Olson	Staff Leader for Field Investigation, USDA-SCS, National Soil Survey Center, Lincoln, Nebraska
L. F. Ratliff	USDA-SCS, National Soil Survey Center, Lincoln, Nebraska
E. M. Rutledge	Professor, Department of Agronomy, University of Arkansas, Fayetteville, Arkansas
Steven A. Scheinost	Soil Scientist, USDA-SCS, Lincoln, Nebraska
G. W. Schellentrager	State Soil Scientist, USDA-SCS, Des Moines, Iowa
S. Simkins	Department of Plant and Soil Sciences, University of Massachusetts, Amherst, Massachusetts
Dan R. Upchurch	Soil Physicist, USDA-ARS, Lubbock, Texas
P. L. M. Veneman	Professor of Soil Sciences, Department of Plant and Soil Sciences, University of Massachusetts, Amherst, Massachusetts
R. J. Wagenet	Professor of Soil Science, Department of Soil, Crop, and Atmospheric Sciences, Cornell University, Ithaca, New York
L. P. Wilding	Professor of Soil and Crop Sciences, Department of Soil and Crop Sciences, Texas A&M University, College Station, Texas
Fred J. Young	Soil Survey Project Leader, USDA-SCS, Columbia, Missouri

Conversion Factors for SI and non-SI Units

Conversion Factors for SI and non-SI Units

To convert Column 1 into Column 2, multiply by	Column 1 SI Unit	Column 2 non-SI Unit	To convert Column 2 into Column 1, multiply by
		Length	
0.621	kilometer, km (10^3 m)	mile, mi	1.609
1.094	meter, m	yard, yd	0.914
3.28	meter, m	foot, ft	0.304
1.0	micrometer, μm (10^{-6} m)	micron, μ	1.0
3.94×10^{-2}	millimeter, mm (10^{-3} m)	inch, in	25.4
10	nanometer, nm (10^{-9} m)	Angstrom, Å	0.1
		Area	
2.47	hectare, ha	acre	0.405
247	square kilometer, km^2 (10^3 m)^2	acre	4.05×10^{-3}
0.386	square kilometer, km^2 (10^3 m)^2	square mile, mi^2	2.590
2.47×10^{-4}	square meter, m^2	acre	4.05×10^3
10.76	square meter, m^2	square foot, ft^2	9.29×10^{-2}
1.55×10^{-3}	square millimeter, mm^2 (10^{-3} m)^2	square inch, in^2	645
		Volume	
9.73×10^{-3}	cubic meter, m^3	acre-inch	102.8
35.3	cubic meter, m^3	cubic foot, ft^3	2.83×10^{-2}
6.10×10^4	cubic meter, m^3	cubic inch, in^3	1.64×10^{-5}
2.84×10^{-2}	liter, L (10^{-3} m^3)	bushel, bu	35.24
1.057	liter, L (10^{-3} m^3)	quart (liquid), qt	0.946
3.53×10^{-2}	liter, L (10^{-3} m^3)	cubic foot, ft^3	28.3
0.265	liter, L (10^{-3} m^3)	gallon	3.78
33.78	liter, L (10^{-3} m^3)	ounce (fluid), oz	2.96×10^{-2}
2.11	liter, L (10^{-3} m^3)	pint (fluid), pt	0.473

CONVERSION FACTORS FOR SI AND NON-SI UNITS

To convert Column 1 into Column 2, multiply by	Column 1 SI Unit	Column 2 non-SI Unit	To convert Column 2 into Column 1, multiply by

Mass

2.20×10^{-3}	gram, g (10^{-3} kg)	pound, lb	454
3.52×10^{-2}	gram, g (10^{-3} kg)	ounce (avdp), oz	28.4
2.205	kilogram, kg	pound, lb	0.454
0.01	kilogram, kg	quintal (metric), q	100
1.10×10^{-3}	kilogram, kg	ton (2000 lb), ton	907
1.102	megagram, Mg (tonne)	ton (U.S.), ton	0.907
1.102	tonne, t	ton (U.S.), ton	0.907

Yield and Rate

0.893	kilogram per hectare, kg ha^{-1}	pound per acre, lb acre^{-1}	1.12
7.77×10^{-2}	kilogram per cubic meter, kg m^{-3}	pound per bushel, bu^{-1}	12.87
1.49×10^{-2}	kilogram per hectare, kg ha^{-1}	bushel per acre, 60 lb	67.19
1.59×10^{-2}	kilogram per hectare, kg ha^{-1}	bushel per acre, 56 lb	62.71
1.86×10^{-2}	kilogram per hectare, kg ha^{-1}	bushel per acre, 48 lb	53.75
0.107	liter per hectare, L ha^{-1}	gallon per acre	9.35
893	tonnes per hectare, t ha^{-1}	pound per acre, lb acre^{-1}	1.12×10^{-3}
893	megagram per hectare, Mg ha^{-1}	pound per acre, lb acre^{-1}	1.12×10^{-3}
0.446	megagram per hectare, Mg ha^{-1}	ton (2000 lb) per acre, ton acre^{-1}	2.24
2.24	meter per second, m s^{-1}	mile per hour	0.447

Specific Surface

10	square meter per kilogram, m^2 kg^{-1}	square centimeter per gram, cm^2 g^{-1}	0.1
1000	square meter per kilogram, m^2 kg^{-1}	square millimeter per gram, mm^2 g^{-1}	0.001

Pressure

9.90	megapascal, MPa (10^6 Pa)	atmosphere	0.101
10	megapascal, MPa (10^6 Pa)	bar	0.1
1.00	megagram per cubic meter, Mg m^{-3}	gram per cubic centimeter, g cm^{-3}	1.00
2.09×10^{-2}	pascal, Pa	pound per square foot, lb ft^{-2}	47.9
1.45×10^{-4}	pascal, Pa	pound per square inch, lb in^{-2}	6.90×10^3

(continued on next page)

Conversion Factors for SI and non-SI Units

To convert Column 1 into Column 2, multiply by	Column 1 SI Unit	Column 2 non-SI Unit	To convert Column 2 into Column 1, multiply by
		Temperature	
$1.00\ (K - 273)$	Kelvin, K	Celsius, °C	$1.00\ (°C + 273)$
$(9/5\ °C) + 32$	Celsius, °C	Fahrenheit, °F	$5/9\ (°F - 32)$
		Energy, Work, Quantity of Heat	
9.52×10^{-4}	joule, J	British thermal unit, Btu	1.05×10^{3}
0.239	joule, J	calorie, cal	4.19
10^{7}	joule, J	erg	10^{-7}
0.735	joule, J	foot-pound	1.36
2.387×10^{-5}	joule per square meter, J m^{-2}	calorie per square centimeter (langley)	4.19×10^{4}
10^{5}	newton, N	dyne	10^{-5}
1.43×10^{-3}	watt per square meter, W m^{-2}	calorie per square centimeter minute (irradiance), cal cm^{-2} min^{-1}	698
		Transpiration and Photosynthesis	
3.60×10^{-2}	milligram per square meter second, mg m^{-2} s^{-1}	gram per square decimeter hour, g dm^{-2} h^{-1}	27.8
5.56×10^{-3}	milligram (H$_2$O) per square meter second, mg m^{-2} s^{-1}	micromole (H$_2$O) per square centimeter second, µmol cm^{-2} s^{-1}	180
10^{-4}	milligram per square meter second, mg m^{-2} s^{-1}	milligram per square centimeter second, mg cm^{-2} s^{-1}	10^{4}
35.97	milligram per square meter second, mg m^{-2} s^{-1}	milligram per square decimeter hour, mg dm^{-2} h^{-1}	2.78×10^{-2}
		Plane Angle	
57.3	radian, rad	degrees (angle), °	1.75×10^{-2}

CONVERSION FACTORS FOR SI AND NON-SI UNITS

Electrical Conductivity, Electricity, and Magnetism

	Column SI	Column non-SI	
10	siemen per meter, S m^{-1}	millimho per centimeter, mmho cm^{-1}	0.1
10^4	tesla, T	gauss, G	10^{-4}

Water Measurement

9.73×10^{-3}	cubic meter, m^3	acre-inches, acre-in	102.8
9.81×10^{-3}	cubic meter per hour, m^3 h^{-1}	cubic feet per second, ft^3 s^{-1}	101.9
4.40	cubic meter per hour, m^3 h^{-1}	U.S. gallons per minute, gal min^{-1}	0.227
8.11	hectare-meters, ha-m	acre-feet, acre-ft	0.123
97.28	hectare-meters, ha-m	acre-inches, acre-in	1.03×10^{-2}
8.1×10^{-2}	hectare-centimeters, ha-cm	acre-feet, acre-ft	12.33

Concentrations

1	centimole per kilogram, cmol kg^{-1} (ion exchange capacity)	milliequivalents per 100 grams, meq 100 g^{-1}	1
0.1	gram per kilogram, g kg^{-1}	percent, %	10
1	milligram per kilogram, mg kg^{-1}	parts per million, ppm	1

Radioactivity

2.7×10^{-11}	becquerel, Bq	curie, Ci	3.7×10^{10}
2.7×10^{-2}	becquerel per kilogram, Bq kg^{-1}	picocurie per gram, pCi g^{-1}	37
100	gray, Gy (absorbed dose)	rad, rd	0.01
100	sievert, Sv (equivalent dose)	rem (roentgen equivalent man)	0.01

Plant Nutrient Conversion

	Elemental	Oxide	
2.29	P	P$_2$O$_5$	0.437
1.20	K	K$_2$O	0.830
1.39	Ca	CaO	0.715
1.66	Mg	MgO	0.602

1 The Need to Quantify Spatial Variability

R. W. Arnold

USDA-SCS
Washington, District of Columbia

L. P. Wilding

Texas A&M University
College Station, Texas

ABSTRACT

Never before in the history of soil science has the knowledge of soil spatial variability been so germane. It reaches to the heart of the pedology profession and is critical to the success of agronomic practice, agricultural development, land management, and earth science on a global scale. One of the continuing challenges for pedologists and allied earth scientists is to develop integrated system models to scale spatial knowledge of soils from microsamples to pedons, landforms and the pedosphere. Quantification of the magnitude, location and causes of spatial variability is an essential but insufficient ingredient of soil surveys. The final payoff is to communicate this knowledge to user clientele in flexible formats that provide for probability risk assessments and alternative land-use decisions. In this electronics era of exploding information systems, we have an opportunity as never before to (i) impact the direction of pedology, (ii) quantify spatial knowledge, (iii) add quality to soil resource inventories, and (iv) test multiple working hypotheses. Real-time assessment of spatial variability allows differential application and treatment of chemicals, pesticides, irrigation waters, and waste products on small site-specific areas. This chapter elucidates the traditions, relevance, challenges, and opportunities for pedologists to assume leadership roles in augmenting spatial knowledge and its application for wise stewardship of soil resources, the sustainability of the global environment, and the preservation of present and future civilizations.

In 1972, academician Gerasimov prefaced a report (Fridland, 1976, p. vii) on the patterns of soil cover, a subject he referred to as pedogeography, with these remarks, "The overall status of the problem is patently unsatisfactory and stimulates further scientific endeavors, since knowledge of this cycle of spatial-genetic regularities significantly deepens our understanding of the evolutionary history and present day habit of the soil cover and the genesis, properties, and regimes of soils...."

He went on to explain that the study of pedogeography permitted more correct mapping and generalization of the soil patterns, improved appraisal of land resources, and of the predictions of soil modifications under natural conditions as well as under the influence of human economic activities.

Fridland (1976) outlined a 10-category scheme of characterizing the typology of simple geographic soil combinations. The higher categories involved the nature of the relations and degree of contrast among components and the lower categories were based on the degree of dissection, and shape of elementary areas and finally on the nature of the boundaries between the components.

A few years earlier Kellogg (1966) discussed the use of soil surveys for community planning and made the point that to interpret the results of physical, chemical, and biological studies of soil samples one had to return to the soil outdoors and integrate the results with all the other unsampled properties and processes. In fact he summarized his discussion with these words, " . . . our interest in any soil is not primarily how it is now or how it will function in the natural landscape but how it will respond to treatment and manipulation."

After studying the effect of size and shape of soil map units in determining soil-use potential, Oschwald (1966) concluded that the challenge for survey interpretations was "to more adequately synthesize the component parts of soil landscapes into segments of space that are meaningful to the soil user in relation to the soil-use decisions he faces."

Thus, we recognize that soil diversity and variability have been of importance in soil survey for many years and that we are presenting a progress report.

WHY BOTHER WITH SOIL VARIABILITY?

We strongly believe that soil scientists and others who deal with soil variability can make valuable contributions toward the evaluation and beneficial use of soil resources in the future. The opportunities as well as the challenges go far beyond creating awareness. They extend deep into the quantification of our knowledge and skills and reach to the heart of being able to effectively communicate with those who can apply such information.

Professionals have the skills, knowledge, and expertise to bring their talents to bear on problems that involve their discipline. The trained and experienced soil scientists are the professionals who have the responsibility to care and share in the future of our planet. If we don't accept these challenges, others who are less qualified will; and soil scientists will be displaced by apathy. Of all the disciplines representing earth science, pedologists have more knowledge of the upper 2 m of the earth's surface than all the rest and should claim this zone as their field of expertise.

We further believe that meaningful responses and decisions are based on quality information. Quality models, quality soil maps, quality descriptions, quality analytical data, and quality interpretations based on relevant criteria of evaluation are products of our trade. Of special importance is quality technical service provided by quality soil scientists.

Pedologists have long recognized spatial variability as the lifeline of their professions; but many earlier soil surveys did not adequately convey to users the complexity of soil property and soil composition variability that comprises landscape units (Becket & Webster, 1971; Miller, 1978; Wilding & Drees, 1983; Mokma, 1987; Wilding, 1988). The credibility of interpretations from soil surveys is heavily weighted on developing and maintaining a quality-assurance control program.

It has almost become axiomatic that if you don't have the quality; the quantity doesn't matter. We hear a lot about quality control and quality assurance both of which are activities to ensure the reliability of our products and services. Mapping acres and acres of soil maps without accuracy and without quality are not in our best interest as professional soil scientists or of pedology as a scientific discipline. Documenting reality to "be sure" will always be good business and sound science.

OUR TRADITION

The foundation of soil survey and related activities (pedology) is the degree of certainty with which soil variability is associated with scales of time and space. The factors of soil formation (parent material, biota, climate, topography, and time) are themselves variable in time and space and at any one moment we observe a complex mosaic on the earth's surface. Interactions among the factors are thought of as processes. They vary according to their own time scales and are constrained in space that results in intricacies far beyond simple descriptions of cause and effect.

The National Cooperative Soil Survey (NCSS) deals with soil patterns in space and in time. We have said that if soil variability, either spatial or temporal, is systematic then we can map it. However, if the variability is random we can only describe it because its location is not predictable, only its occurrence. We have massive amounts of relational observations and analytical solutions hidden from sight — casually tucked away in the recesses of our minds or in files at the office somewhere. All too often this information is only loosely connected to the products delivered.

We believe that the strength of this foundation is capable of supporting a superstructure as yet undefined and perhaps not yet imagined. The reliabilities of observed relationships are often very good; however, they can be improved with additional documentation, verification, and communication. Improved methods of observing and new techniques of discovery will permit us to build on the past and reach for the future.

TODAY'S RELEVANCE

Sophisticated users of soil information in the USA are now applying high technology to obtain a better combination of precious resources. Computer chips imprinted with a soil map and crop conditions are used to adjust application rates of fertilizers, seeds, pesticides, and even irrigation water. Geographic location and site specificity are here today — the possibilities to refine resource combinations are fantastic!

There is a continuing recognition of values associated with soils and their combinations in space: tax assessment, land values, route locations, preservation of areas deemed important for society such as fragile land, wilderness, prime farm land, and wetlands. Soil maps and the associated attribute information are vital to the identification, inventory, and evaluation processes underlying policy decisions concerning land uses.

The degree of uncertainty associated with time and space can be reduced by improved documentation of field variability. Wilding (1985) and Wilding and Drees (1978, 1983) mention that such documentation improves the estimates of central tendency and variance for specific classes, helps to quantify pedogenesis, and permits the composition of map units to be verified. They also suggested that the study of spatial variability assists in developing sampling designs and statistical models for soil survey and other pedogenic applications, and improves optimal allocation of sampling units for efficient statistical designs. Characterization of spatial patterns should also assist in separating systematic from random error in the analysis of landforms; and three-dimensional representations enable us to easily visualize soil behavior and pedogenesis.

LEARNING FROM ABROAD

When members of the NCSS go overseas to assist others to help themselves it often raises old questions to new levels of significance. The USA passed through different stages of agricultural and economic development quickly and many of us are products of only the last stage that involves high technology and massive energy inputs.

Agronomic research for limited resource farmers all too often comes face to face with microvariability of soils that overshadows our usual methods of analysis. Main treatment effects of soil management experiments may be noise confounded by the site variability. There is a need to quantify soil variability and determine the scale or scales of its occurrences if we want to optimize resource allocations. In addition, there is the need to develop explanations of observable variability and to determine the possibilities of prediction. The probabilities of success are often linked with the perception of risk; thus it is a worthy goal to minimize risks associated with soil resources.

Even when site variability swamps our traditional ways of dealing with it with classical statistical approaches (ignoring or wiping out such differences by blocking treatments) there remains the question, How extensive are the areas that contain these levels of microvariability? Cropping sequences, farming systems, conservation practices, and techniques of renovation of degraded soils may be unduly influenced by these patterns of nature. What if there is a nesting effect of significance to the use of resources that can be partially solved by quantifying soil variability? Would you risk the chances of not trying to learn?

Among the new questions for old data is one about water management and the usefulness of subwatersheds. Do these small hydrologic response areas provide new insight into water and soil management that are more useful than our traditional measures of variability? How does water behave in a microwatershed? How strong are the connections and responses throughout the area? Can such questions generated from research in developing countries provide fresh insights into management techniques at home?

In many developing countries the management of degraded soil resources is all too often a necessity due to social and economic pressures. The luxury of a "Conservation Reserve" or of "Set Aside Acres" practiced in developed countries with excess food supplies simply doesn't exist everywhere. Where an environment has been degraded by erosion, desertification, salinization, waterlogging, or pollution; what are satisfactory ways to stabilize the conditions, conserve and manage the remaining resources, and sustain agricultural production for growing populations? Tough options for which we are ill-prepared to deal with adequately; however, education, common sense, and scientific integrity are fundamental building blocks for the right solutions and for responsible decision-making. Yes, we now realize that the full spectrum of stages of agricultural development still exists and that each stage must be addressed by someone, somewhere.

EXPANDING OUR CONCEPTS

"The problem of global environmental change is critical and urgent. For scientists, understanding the changes now in progress — both natural and anthropogenic — and predicting their future courses are unprecedented challenges," as stated by members of the U.S. National Committee for the International Geosphere-Biosphere Program (IGBP) (Committee on Global Change, 1988). They point out the fundamental paradigm is that the prediction and ultimate management of environmental problems inescapably require development of a new earth system science aimed to improve understanding of the earth as an integrated whole.

The stated objective of the IGBP is "to describe and understand the interactive physical, chemical, and biological processes that regulate the total earth system, the unique environment that it provides for life, the changes that are occurring in this system, and the manner in which they are influenced by human activities" (Committee on Global Change, 1988, p. 2).

The U.S. Committee has recently recommended five research initiatives, namely: (i) water–energy–vegetation interactions, (ii) fluxes of materials between terrestrial ecosystems and the atmosphere and ocean, (iii) biogeochemical dynamics of ocean interactions with climate, (iv) earth system history and modeling, and (v) human interactions with global change. In as much as soils are sinks, sources, and filtering membranes, as well as being blocks of memory, it appears to us that soil science can be an important contributor to at least four of these areas of research.

For example, when climatic belts shift, for whatever causes, the biota respond more quickly than the soil because soils are fixed spatially. What relationships have we observed, or learned about, or simulated, which will help governments and communities make adjustments that are in keeping with an integrated world and its environment?

For example, we are aware that as we enlarge scales we can recognize, identify and delineate more detail than at the smaller scales. Is there a theory or set of concepts that link the levels of variability together throughout the continuum of space? Is there a self-similarity that overrides the concepts we impose on our understanding of detailed soil surveys? Is the similarity that we search for in spatial variability only present in the degree of complexity of shapes and sizes and not in the organizing patterns of apparent chaos?

Global issues go far beyond current levels of understanding and beyond our present efforts to cooperate and integrate information into models of earth system processes. There truly are unprecedented challenges ahead.

THE FUTURE IS HERE AND NOW

In this electronic era there are many options for transmitting what we already know. There appear to be as many options as there are innovative imaginations to assist those who want, or will use, the information that can be provided. It also seems to suggest that specialists other than soil scientists can and will become involved in the packaging and transmission of soil-related information. It is incumbent upon us to design and build soil databases that are flexible and capable of responding to the multipurpose inquiries of future consumers.

It is time to re-evaluate what we already know relative to soil resources. Throughout the admonitions engendered by the global-change paradigm is a thoughtful thread of caution. It reminds us to be aware of and listen to the unusual, the catastrophic, the strange frequency of events or rates that have and are occurring. The disturbing data points that reduce correspondence may hold the keys that open doors to new and more exciting explanations than we imagined. The possibilities of being surprised should be welcomed rather than hidden or shunned.

Now is the time to examine and record the basis of our observations and their relationships; and bring together the explanations and the experimental results that inform us of soil genesis, soil distribution patterns, and soil behavior. Quantification of what we feel and know is the basis for our actions. The utility of our information depends on the degree of its certainty. Is our information quantified?

Each of us in our own way should explore the areas of our limited understanding in this era of knowledge and information explosion. There are theories and concepts and techniques employed by other disciplines to comprehend and explain information, but by convention or ignorance we have not availed ourselves of this cutting edge of knowledge. Information specialists also deal with the collection of data, the production of information, and the flow of information, from place to place and person to person, but we have utilized their expertise sparingly.

Included in our collective areas of ignorance are:

1. Risk assessment, risk avoidance, and risk management as it relates to soil resources and their variability.
2. Space and time statistics and their interactions associated with soils, and
3. Experiential knowledge of indigenous populations concerning use and management of soil resources.

SO WHAT?

Soil survey is a predictable study of soils as geographic bodies, by whatever manner they are conceived; and it determines the unique relationships of sets of soil properties that are observed in nature (Soil Survey Staff, 1975).

Arnold (1983) confirmed that because of the art and scale used in map making, and the recognition of intermingled soil bodies having contrasting qualities, delineating areas containing the same limits of variability as taxonomic classes is precluded.

We, ourselves, are part of the problems that will be evident in the following chapters. However, we also are part of the solutions to these problems. Obviously we should, can, and will be involved because we are soil scientists contributing to the National Cooperative Soil Survey Program. Our mission is to help people understand and use wisely the soil resources not only of this country but of the world.

It is imperative that we reach out to help others. It is the combined search for truth and understanding that will lead us to awareness and comprehension and responsible actions to make this a sustainable global environment. For us it is the search for, and the discovery of, new dimensions of our very existence.

REFERENCES

Arnold, R.W. 1983. Concepts of soils and pedology. p. 1–21. *In* L.P. Wilding et al. (ed.) Pedogenesis and soil taxonomy. I. Concepts and interactions. Developmentts in Soil Science 11A. Elsevier, Amsterdam.

Beckett, P.H.T., and R. Webster. 1971. Soil variability (a review). Soils Fert. 34:1–15.

Committee on Global Change. 1988. Toward an understanding of global change. National Academy Press, Washington, DC.

Fridland, V.M. 1976. Pattern of the soil cover. Preface. Keter Publ. House, Jerusalem.

Kellogg, C.E. 1966. Soil surveys for community planning. p. 1-7. *In* L.J. Bartelli et al. (ed.) Soil surveys and land use planning. ASA, SSSA, Madison, WI.

Miller, F.P. 1978. Soil surveys under pressure: The Maryland experience. J. Soil Water Conserv. 33:104-111.

Mokma, D.L. 1987. Soil variability of five landforms in Michigan. Soil Surv. Land Eval. 7:35-51.

Oschwald, W.R. 1966. Quantitative aspects of soil survey interpretation in appraisal of soil productivity. p. 152-159. *In* L.J. Bartelli et al. (ed.) Soil surveys and land use planning. ASA, SSSA, Madison, WI.

Soil Survey Staff. 1975. Soil taxonomy. A basic system of soil classification for making and interpreting soil surveys. USDA-SCS Agric. Handbk. 436. U.S. Gov. Print. Office, Washington, DC.

Wilding, L.P. 1985. Spatial variability: Its documentation accommodation and implications to soil surveys. p. 166-194. *In* D.R. Nielsen and J. Bouma (ed.) Soil spatial variability. Proc. Workshop ISSS and SSSA, Las Vegas, NV. 30 Nov.-1 Dec. 1984. PUDOC, Wageningen, Netherlands.

Wilding, L.P. 1988. Improving our understanding of the composition of soil-landscape. p. 13-35. *In* H.R. Finney (ed.) Proc. Int. Interactive Workshop on Soil Resources: Their Inventory, Analysis, and Interpretation for Use in the 1990's, St. Paul, MN. 22-24 Mar. 1988. Univ. of Minnesota Coop. Ext. Serv., St. Paul, MN.

Wilding, L.P., and L.R. Drees. 1978. Spatial variability: A pedologist's viewpoint. p. 1-12. *In* M. Drosdoff et al. (ed.) Diversity of soils in the tropics. ASA Spec. Publ. 34. ASA, SSSA, Madison, WI.

Wilding, L.P., and L.R. Drees. 1983. Spatial variability and pedology. p. 83-116. *In* L.P. Wilding et al. (ed.) Pedogenesis and soil taxonomy. I. Concepts and interactions. Developments in Soil Science 11A. Elsevier, Amsterdam.

2 Predicting Variability of Soils from Landscape Models

G. F. Hall
Ohio State University
Columbus, Ohio

C. G. Olson
USDA-SCS
Lincoln, Nebraska

ABSTRACT

Historically, use of landscape models has shown that landscapes are predictable; they have a large nonrandom variability component. This nonrandom variability can be used to predict soils on the landscape if the methodology both to describe and to quantify processes that govern landscape development is understood. The bases for our understanding of soils and landscapes are the concepts of Davis and Penck and those of G. Milne, L.C. King and R.V. Ruhe in whose models process plays an important role. The landscape must be defined in three dimensions, the lateral as well as the vertical changes in stratigraphic materials. The hydrologic characteristics of the system, particularly lateral flow, must be determined. A better understanding of stratigraphic control on hydrologic parameters is needed. Neither landscape nor process components are defined adequately by present soil map units. Milne's catena model is an excellent foundation for integration of these concepts.

Many landscape components are predictable. The soils associated with these components should also be predictable. There are two categories of variability in most landscape studies — systematic and random. In which category one places a set of parameters depends on the level of observation. As the spacing of observations decreases, the apparent randomness of the landscape decreases. As pointed out by Wilding and Drees (1983, p. 85), in a discussion of soil variability, if enough observations are made, the error in predicting the landscape variability can be understood and explained.

In most cases sufficient information is not available to understand the processes that shape present landforms. Most attempts at modeling landscapes have been unsuccessful because the landscape was either looked at in too little detail or the landscape was considered only in two dimensions. A model is useful only if it represents the actual field situation. In soil-landscaping modeling, the model must represent the processes that actually occur in the field. Because most of the processes related to the evolution of the landscape act in three dimensions, the two-dimensional approach overlooks the full impact of many processes. Models must evolve as the result of three-dimensional considerations to fully understand the processes and morphology of the landscape. It follows that the three-dimensional view of the relation between soils and landscapes must be understood before these models can be developed and implemented.

LANDSCAPE TERMINOLOGY

Background

A common terminology is required for the description of landscapes and landscape evolution. The basis for our present terminology and understanding of soils and landscape models lies in the historic concepts of William Morris Davis and Walther Penck. The now classic "geographic cycle" (Davis, 1899) was a time-dependent landscape evolutionary model. Landscapes progressed through a series of stages from youth through maturity to old age. Slope reduction was by downwearing. Differences in geology became insignificant with time. Process was largely ignored in favor of an emphasis on the form of the land.

Penck (1924), among the first to break from the Davisian school, emphasized backwearing with parallel slope retreat in his landscape model. Broad concave slopes extended away from drainageways at the expense of uneroded interfluves. The Davis and Penck theories were general landscape models and did not provide details of hillslope development. Wood (1942) and King (1953) formulated models for the fully developed hillslope. Their models were followed closely by the soil-geomorphic model of R.V. Ruhe (1956, 1960) in which soil properties were integrated with hillslope models.

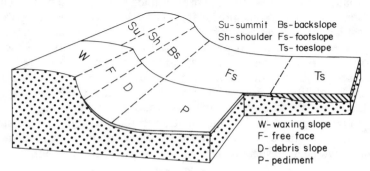

Fig. 2-1. The elements of a hillslope model with summit shoulder backslope, footslope, and toeslope. Modified by Ruhe (1960) from earlier models by Wood (1942) and King (1953) in the foreground (from Ruhe, 1960).

Ruhe (1960, 1975) modified the Wood and King model by proposing the hillslope elements of summit, shoulder, backslope, footslope, and toeslope (Fig. 2-1). With minor modifications, these elements are now widely used for the description of hillslope positions. Following the introduction of the hillslope profile elements, Ruhe formalized terms of geomorphic slope components (Ruhe & Walker, 1968; Ruhe, 1975): head slope for the concave portion, side slope for the linear portion, and nose slope for the convex portion (Fig. 2-2).

The Catena Concept

Milne was an early contributor to the idea that soils are uniquely related to landscape position with his introduction of the catena concept (Milne, 1936a,b). The concept was derived from his East African work involving a range of landscape types and soils. His was the first real process-oriented approach to the study of landscapes. He used the term catena to suggest that the soils on the landscape are related as links in a chain. Processes that occur on one part of a given landscape influence not only the soils at that location but soils on other parts of the landscape: "Soil differences were then brought about by drainage conditions, differential transport of eroded material, and leaching, translocation, and redeposition of mobile chemical constituents" (Milne, 1936a). From this statement it is clear that in Milne's model of a soil landscape, pedogenic processes in concert with hydrologic properties of the materials encompass the entire soil landscape and are not restricted to the soil at just one point on that landscape. His model is an ideal approach to the study of hillslopes, integrating the geomorphic processes of erosion, transport and deposition, and water movement with pedogenic processes.

Milne's catena concept had two variants: (i) all soils of a catena are formed in a single material, and (ii) soils of a catena may be formed in two or more materials. Soils of a catena differ in the first of these because of (i) drainage conditions, (ii) differential transport and deposition of eroded material, and (iii) leaching, translocation and redeposition of mobile chemical constituents. In the second variant, a geologic factor of multiple-parent materials is added.

The catena model proposed by Milne evolved into the current limiting toposequence model. The toposequence, as now commonly used in soil survey, is in practice a hydrosequence using primarily soil profile colors as indicators of water table elevation differences. Water movement, particularly lateral movement, and the properties related to the movement of materials in solution and suspension are usually not addressed in this model. Species and concentrations of materials translocated through the landscape are commonly ignored.

Fridland (1972), studying the soil cover in the USSR, also emphasized the importance of lateral transport and related processes in understanding the genetic interrelationship of soils. He believed that it was inconclusive to make inferences about soil genesis from the study of a single soil profile because this assumed the movement of substances to be primarily vertical.

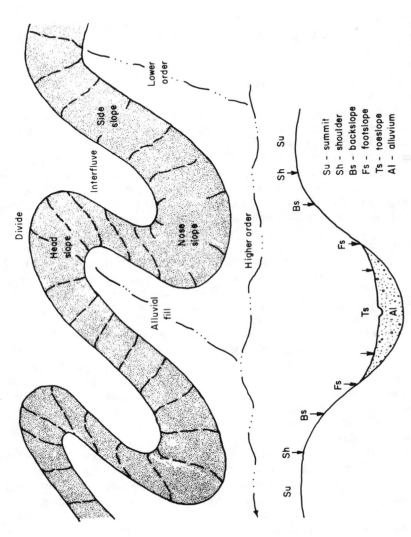

Fig. 2-2. The geomorphic components of an incised valley illustrating head slope, side slope, and nose slope. Lower diagram is a hillslope profile between two interfluves (from Ruhe and Walker, 1968).

SOIL VARIABILITY AND LANDSCAPE MODELS

Young (1972, 1976) identified static and dynamic causes for catenary differentiation while acknowledging that most slopes have an interplay of both. Static causes were governed by site differences such as slope angle and depth to the water table. Dynamic causes were governed by position on the slope and included primarily downslope transport processes.

LANDSCAPE MORPHOLOGY

In order to properly evaluate the landscape and the relationship of the soils to that landscape, the morphology of the landscape must be determined. This evaluation should include at a minimum: gradient, aspect, and vertical and horizontal curvature.

Slope Curvature

An important determinant of water movement and the resultant geomorphic and pedologic processes is the planar curvature of the slopes, both the longitudinal and latitudinal profile. One of the first to consider the importance of curvature was Aandahl (1948). In this early work slope curvature was related to differences in fertility status and soil morphological properties. Troeh (1964) illustrated the possible combinations of vertical and horizontal slope configuration and combined them into four basic convex-concave combinations. Ruhe (1975) described slope curvature using three components: (i) slope gradient, (ii) slope length, and (iii) slope width. Changes in curvature could be represented by a matrix of nine basic forms. Huggett (1975) added surface flow lines to the basic slope shapes (Fig. 2-3).

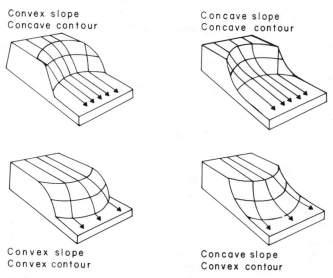

Fig. 2-3. Four basic slope shapes with surface flow lines (from Huggett, 1975).

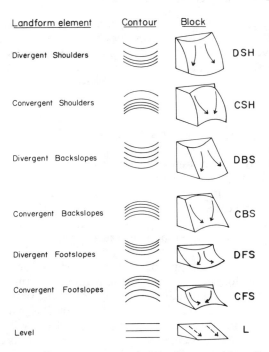

Fig. 2-4. Hillside elements and curvature on seven landscape positions (from Pennock et al., 1987).

In this illustration runoff and throughflow converge in situations where there is a concave contour and diverge where there is a convex contour. The concavity or convexity of the slope modifies the erosive power of surface-water flow and influences the path of water movement through the soil. The soil tends to become saturated and seepage occurs in the footslope and head slope positions where the slope is concave. Convex shoulder elements and nose positions are drier due to divergence of moisture in these hillslope positions. Thus, convexity and concavity of slopes are important controls on water movement and directly related to the variability of soils on a hillslope.

In their work on glacial till soils in Canada, Pennock and others (Pennock et al., 1987) combined the hillslope elements and curvature to identify seven different hillslope positions (Fig. 2-4). They considered these positions to have different processes and potential for different soil properties, thus different crop yields. The hillslope positions were: (i) divergent shoulder, (ii) convergent shoulder, (iii) divergent backslope, (iv) convergent backslope, (v) divergent footslope, (vi) convergent footslope, and (vii) level where both the profile and the contour were linear. In all cases the water is assumed to be moving along the maximum slope gradient and normal to the contours.

Figure 2-5 illustrates one combination of convergent and divergent characteristics that cause a complex pattern of water movement in a hillslope system (Pennock et al., 1987). Of the landform elements, shoulder, backslope,

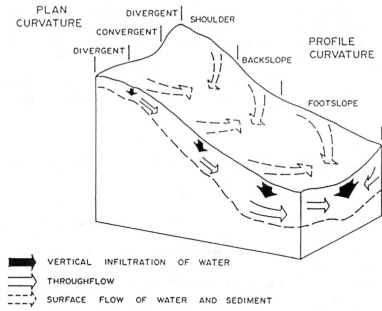

Fig. 2-5. Pattern of moisture movement in a complex hillslope system (from Pennock et al., 1987).

and footslope, the element with the least moisture available for pedogenic development would be the divergent shoulder and the element with the most moisture would be the convergent footslope. The other elements would be expected to have intermediate moisture contents. Because the movement and accumulation of water on these hillslope positions differ, the resultant soils would be expected to be different. In the field, Pennock et al. (1987) found that the relationship between moisture content and landform elements was shoulder < backslope < footslope. In evaluating the A horizon thickness and depth to carbonates they found the relationship shoulder < backslope < level < footslope. In later work Pennock and Acton (1989) utilized the hillslope elements to assign a relative contribution of sedimentation and hydrologic activity. The soil characteristics they observed in the field were in accord with the water movement in the landscape.

Landscape Morphology Analysis

Using standard USGS quadrangle maps Lanyon and Hall (1983a,b) determined the gradient, aspect, and vertical and horizontal curvature of a small watershed in eastern Ohio. The geologic material in the watershed consisted of a complex pattern of residuum, colluvium and loess. Maps of these morphological characteristics were computer generated (Fig. 2-6 and 2-7). In the field the geologic materials and the solar radiation were determined for various positions in the watershed. The field data, together with the slope curvature, were used to generate a map of predicted landscape instability in the watershed (Fig. 2-8). These areas had both horizontal and vertical concavity and thus had the maximum accumulation of moisture. The least stable areas were identified as having geologic material considered susceptible to slippage. These predicted unstable sites were on steep north-facing slopes that received lesser amounts of solar radiation. Field checking located five sites with current soil mass movement or evidence of relict landslips that coincided with those areas of predicted instability (Fig. 2-8).

WATER MOVEMENT AND LANDSCAPE DEVELOPMENT

Flow Paths

Water movement is one of the most important driving forces in the processes involved in soil genesis and landscape evolution. Locating flow paths is important for both the reactions that cause pedogenic variability and the many practical aspects of soil behavior.

Figure 2-9 (Gerrard, 1981) illustrates the major flow paths for water movement on a hillslope: (i) surface runoff, (ii) infiltration, (iii) throughflow, and (iv) groundwater recharge. Water moves across the surface as saturated flow until it enters a stream. By this path the water can create severe erosional problems, removing surficial material from a part of the landscape and either depositing the sediment locally or carrying it away from the hillslope. Water

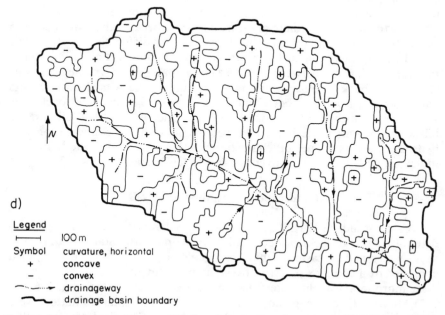

Fig. 2-6. Map showing the horizontal land-surface curvature in an eastern Ohio drainage basin (from Lanyon & Hall, 1983a).

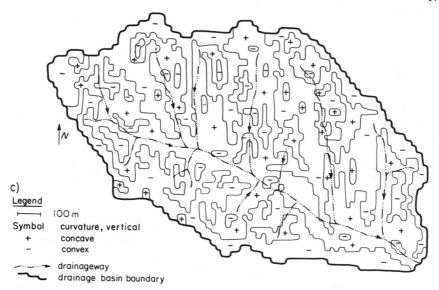

Fig. 2-7. Map showing the vertical land-surface curvature in an eastern Ohio drainage basin (from Lanyon & Hall, 1983a).

also may enter the soil and percolate through the soil profile. During this process the water carried with it soluble and suspended materials. Much of the suspended material is deposited lower in the soil profile adding to the anisotropy of the soil. In an open system, some suspended material and a larger portion of the soluble material are carried out of the soil profile and enters the groundwater or a stream and are carried out of the local hydrologic system.

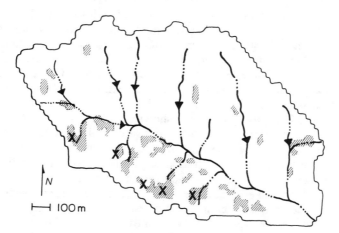

Fig. 2-8. Map showing predicted landscape instability in an eastern Ohio drainage basin (from Lanyon & Hall, 1983b).

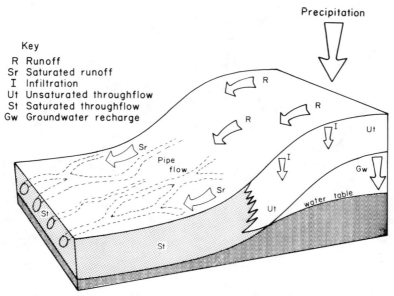

Fig. 2-9. Flow paths for water movement on slopes (adapted from Gerrard, 1981).

Perhaps one of the most important contributors to the variability of soils on the hillslope is the water that moves downward in the soil profile, encounters a less permeable zone or horizon and moves laterally as saturated or unsaturated throughflow or in conduits such as pipes. In some soils the amount of moisture moving laterally exceeds that moving vertically. Soil welding (Ruhe & Olson, 1980) commonly results in lateral flow. The mechanism for a typical developmental sequence of an indurated or partly indurated zone is provided in Olson and Hupp (1986).

Bear et al. (1968) reported that the combined hydraulic conductivity was greater parallel to soil horizons than normal to the horizons. Water moving laterally within the soil carries suspended and soluble material. In some cases eluviation and illuviation occur laterally similar to that which occurs vertically in a soil profile. An example of this is the entrainment of Fe or carbonate in an upslope position followed by their deposition in a lower landscape position. This phenomenon was described by Milne in his African catena work (Milne, 1936b). Throughflow was shown to be unsaturated in the upper, convex and linear portions of the slope and saturated in the lower, concave footslope portion. Whether the flow at a point is saturated or unsaturated is dependent on many things such as precipitation duration and intensity, horizon permeability, slope gradient, and vegetative cover. In many situations the subsurface flow will emerge on the lower footslope or toeslope as a seep. As a result of the various chemical and physical processes that result from the types, directions and quantity of water movement in and on the hillslope, the soils on the hillslope vary in direct relation to these factors.

Soil Anisotropy

To understand the processes involved in the catena concept it is necessary to recognize that all soils are anisotropic. Anisotropy is a major factor in determining the processes acting in the soil, their degree of intensity, and the resulting soil morphology. Anisotropy in soils is the result of pedogenic horizonation, sedimentation, geologic structures, and compaction. Pedogenic horizonation is dominated by movement of mobile constituents either in suspension or solution and subsequent deposition in the soil. The parent materials commonly occur in strata of varying thicknesses and contrasting materials. This results in differential weathering and a lack of uniformity among resulting soils. Soil compaction and increased density can result from naturally occurring pedogenic processes or from cultivation practices. Compaction has become more common with the use of heavy machinery and the introduction of more monocultural farming practices.

One of the most important aspects of soil anisotropy as it relates to soil variability is its influence on the movement of water in the soil. Water entering the soil at the surface moves downward and outward in all directions if the soil horizons are isotropic. The net resultant movement is vertically downward. Upon reaching a soil horizon or zone with differing permeability resulting from differences in soil properties such as texture, density, pore-size distribution, cementation or mineralogy; some of the resultant flow has a tendency to move laterally. The proportion of the lateral movement is dependent upon the magnitude of change in permeability and the slope gradient and undersaturated conditions, the hydraulic gradient that in most cases is proportional to the slope. This lateral movement is a significant cause of variability in some soils.

Lateral movement of moisture in the soil has been demonstrated in mathematical models (Zaslavsky & Rogowski, 1969) and has been observed and measured many times in field situations (Whipkey, 1965, 1969; Weyman, 1973; Whipkey & Kirkby, 1978; Gaskin et al., 1989). It has also been shown that lateral movement can take place on sloping soils even if the soil is isotropic (McCord & Stephens, 1987). Precipitation intensity and duration and incipient moisture conditions also play a major role in horizontal and vertical permeability on a hillslope (Kirkham, 1947; Reeve & Kirkham, 1951; Chorley, 1978).

Subsurface Flow

Processes as conditioned by soil anisotropy contribute an additional source of variability in soils. Measurement of subsurface flow is a difficult task complicated by the necessity to identify this phenomena for a unique hillslope system or watershed.

Milne included subsurface flow in his model of soil–landscape relations but not until later were the processes related to subsurface flow studied in field situations (Hursh & Hoover, 1941; Hursh, 1944; Amerman, 1956;

Hewlett & Hibbert, 1963; Whipkey, 1965, 1969). Ruhe's work concentrated more heavily on the surficial processes related to landscape evolution than on subsurface flow. Workers in Australia and England led by Dalrymple (Dalrymple et al., 1968; Conacher & Dalrymple, 1977) lent their expertise to the study of surface and subsurface processes that are involved in landscape development. They used a nine-unit landscape in contrast to the five-unit model proposed by Ruhe (1960). The model attempted to integrate slope profile components with material and water flow. Huggett (1973, 1975) developed a three-dimensional approach to simulate material flux in an idealized valley basin. More recently flow systems have been examined in conjunction with studies of hydrologic balance in small watershed systems (Dunne, 1978; Eriksson, 1984).

LAND MANAGEMENT IMPLICATIONS

Landscape Elements and Productivity

Water is a limiting factor in plant growth. Therefore, it should be expected that there is an interrelationship between landscape elements, soils, and the productivity of the soil.

Working in Israel, Sinai et al. (1981) studied the relationship of wheat (*Triticum aestivum* L.) yields and wheat grain size to the convexity or concavity of the landscape position. In an area with about 200 mm of rainfall they found the yields and grain size in concave positions significantly different from those in convex positions (Table 2-1). Yields were almost four times greater in concave positions than in convex positions. Size of the grain was 32% larger by weight in the concave positions.

In a study of the interrelation among landscape position, erosion and corn (*Zea mays* L.) yields, Stone et al. (1985) showed a difference in corn yields on different slope positions (Table 2-2). Because the data were collected in two successive years with very different weather patterns, crop responses as related to slope positions were not the same each year. However, it is clear that yields were different on differing landscape positions. A number of other recent studies have confirmed the relationship between crop yields and landscape positions (Miller et al., 1988; Ciha, 1984).

In similar activities a number of researchers have found a relationship among soil physical and chemical properties and hillslope position (Aguilar & Heil, 1988; Campbell, 1978; McCracken et al., 1989). These differences were due to differential moisture movement.

Table 2-1. The influence of slope curvature on wheat yield and grain size in Israel (from Sinai et al., 1981).

Slope position	Wheat yield kg/ha	Grain size g/1000 grains
Concave	1906	47.8
Convex	503	36.3
Relative yield concave/convex	3.79	1.32

Soil Mapping Units

Much effort has been expended on taxonomic classification of soils during the last few years but the importance of proper representation of landscape relations within and between soil mapping units has been virtually ignored. The same mapping unit is often delineated on convex, concave, and linear slopes. This mapping results in the inclusion of areas of moisture accumulation, moisture depletion and uniform moisture flow within a given mapping unit. In some cases mapping units are delineated on different geomorphic surfaces. Clearly these situations greatly increase the variability that can be expected within a given mapping unit. Wherever possible a single soil mapping unit should be limited to a single landform position. This principle was presented by Guy Smith in a discussion about soil mapping, "The identification of a single series in two or three different landscape positions suggests that neither the genetic nor use relationships of the soil have been sufficiently studied" (Forbes, 1986, p. 49).

If it is not possible to delineate smaller areas, the users of the soil maps should be made aware of the range in moisture and soil properties within the mapping units that extend over several hillslope positions. The user should be educated to the fact that there is a great deal of variability with respect to processes and morphology in these broadly defined mapping units. An understanding of some of the simple processes that are occurring on a landscape that influence this variability would help the user to both understand the variability and make more efficient use decisions for a given mapping unit. The present soil series and mapping unit concepts need revision to take into consideration some of the previously discussed landscape parameters.

Table 2-2. The relationship of corn yields to landscape position for five landscape positions in North Carolina (from Stone et al., 1985).

Location	Landscape position				
	Interfluve	Shoulder	Linear slope	Head slope	Foot slope
	Corn yield				
	Mg/ha				
Davidson Co.					
1981	6.3	6.6	6.2	8.7	6.8
1982	6.5	5.3	4.8	7.0	7.0
Rockingham Co.					
1981	3.6	4.6	5.2	4.8	5.6
1982	3.9	4.7	5.5	5.0	5.8
Wake Co.					
1981	2.2	2.9	2.5	3.1	3.0
1982	6.3	6.5	6.4	6.8	6.4

SUMMARY

Parts of the landscape have a nonrandom variability. Soils are an integral part of the surface of the landscape. It follows then that the soils that are dependent upon hillslope form should be predictable. The major limitation to the prediction of landscapes and soils is the lack of understanding of the processes that were involved in their development. A greater emphasis on interdisciplinary research in the closely allied sciences of geomorphology and pedology is needed to adequately address the systematic variability of soils within landscapes.

Landscapes must be defined in three dimensions and the soils developed on these must be studied with the entire landscape in mind. Milne (1936a) in his classic African work set the stage for the study of the landscape and the soils on that landscape as a single interacting unit by integrating landscape evolution with soil genesis. As currently used, the toposequence is a limited portion of the catena concept; it considers soil water as a passive component in the landscape and not as the major driving force in the development of landscapes and soils.

Water movement is a major driving force in the processes that shapes the landscapes and the soils on them. Understanding the hydrological model is imperative for the understanding of landscape and soil evolution. A hydrologic model must recognize the lateral as well as the vertical component of water movement and the movement of water in both the saturated and the unsaturated state. On most landscapes water and its movement, both surface and subsurface, are major agents for variability of soils on that landscape.

REFERENCES

Aandahl, A.R. 1948. The characterization of slope positions and their influence on the total nitrogen content of a few virgin soils in western Iowa. Soil Sci. Soc. Am. Proc. 13:449–454.

Aguilar, A., and R.D. Heil. 1988. Soil organic carbon, nitrogen, and phosphorus quantities in Northern Great Plains rangeland. Soil Sci. Soc. Am. J. 52:1076–1081.

Amerman, C.R. 1956. The use of unit-source watershed data for runoff prediction. Water Res. 1:499–508.

Bear, J., D. Zaslavsky, and S. Irmay. 1968. Physical principles of water percolation and seepage. UNESCO, Paris.

Campbell, J.B. 1978. Spatial variation of sand content and pH within single contiguous delineations of two soil mapping units. Soil Sci. Soc. Am. J. 42:460–464.

Chorley, R.J. 1978. The hillslope hydrologic cycle. p. 1–42. *In* M.J. Kirkby (ed.) Hillslope hydrology. John Wiley & Sons, New York.

Ciha, A.J. 1984. Slope position and grain yield of soft white winter wheat. Agron. J. 76:193–196.

Conacher, A.J., and J.B. Dalrymple. 1977. The nine unit landscape model: An approach to pedogeomorphic research. Geoderma 18:1–154.

Dalrymple, J.B., R.J. Blong, and A.J. Conacher. 1968. An hypothetical nine unit landsurface model. Z. Geomorphol. 12:60–76.

Davis, W.M. 1899. The geographical cycle. Geogr. J. 14:481–504.

Dunne, T. 1978. Field studies of hillslope flow processes. p. 227–294. *In* M.J. Kirkby (ed.) Hillslope hydrology. John Wiley & Sons, New York.

Eriksson, E. (ed.). 1984. Hydrochemical balances of freshwater systems. Int. Assoc. Hydrol. Sci. IAHS Press, Wallingford, United Kingdom.

Forbes, T.R. (ed.). 1986. The Guy Smith interviews: Rationale for concepts in soil taxonomy. Soil Manage. Support Services Technical Monograph 11. USDA-SCS, Washington, DC.

Fridland, V.M. 1972. The soil-cover pattern: Problems and methods of investigation. p. 1–31. *In* V.M. Fridland (ed.) Soil combinations and their genesis. Amerind, New Delhi.

Gaskin, J.W., J.F. Dowd, W.L. Nutter, and W.T. Swank. 1989. Vertical and lateral components of soil nutrient flux in a hillslope. J. Environ. Qual. 18:403–410..

Gerrard, A.J. 1981. Soils and landforms. Allen and Unwin, Inc., London.

Hewlett, J.D., and A.R. Hibbert. 1963. Moisture and energy conditions within sloping soil mass during drainage. J. Geophys. Res. 68:1081–1087.

Huggett, R.J. 1973. The theoretical behavior of materials within soil landscape systems. Dep. Geography Occasional Paper 19. Univ. College, London.

Huggett, R.J. 1975. Soil landscape systems: A model of soil genesis. Geoderma 13:1–22.

Hursh, C. 1944. Report of sub-committee on subsurface flow. Trans. Am. Geophys. Union 25:743–746.

Hursh, C., and M.D. Hoover. 1941. Soil profile characteristics pertinent to hydrologic studies in the Southern Appalachians. Soil Sci. Soc. Am. Proc. 6:414–422.

King. L.C. 1953. Canons of landscape evolution. Geol. Soc. Am. Bull. 64:721–752.

Kirkham. D. 1947. Studies of hillside seepage in the Iowan drift area. Soil Sci. Soc. Am. Proc. 12:73–80.

Lanyon, L.E., and G.F. Hall. 1983a. Land-surface morphology: 1. Evaluation of a small drainage basin in eastern Ohio. Soil Sci. 136:291–299.

Lanyon, L.E., and G.F. Hall. 1983b. Land-surface morphology: 2. Predicting potential landscape instability in eastern Ohio. Soil Sci. 136:382–386.

McCord, J.T., and D.B. Stephens. 1987. Lateral moisture flow beneath a sandy hillslope without apparent impeding layer. Hydrol. Proc. 1:225–238.

McCracken, R.J., R.B. Daniels, and W.E. Fulcher. 1989. Undisturbed soils, landscapes, and vegetation in a North Carolina Piedmont virgin forest. Soil Sci. Soc. Am. J. 53:1146–1152.

Miller, M.P., M.J. Singer, and D.R. Nielsen. 1988. Spatial variability of wheat yield and soil properties on complex hills. Soil Sci. Am. J. 52:1133–1141.

Milne, G. 1936a. A provisional soil map of East Africa. Amani Memoirs no. 28. Eastern African Agric. Res. Stn., Tanganyika Territory.

Milne, G. 1936b. Normal erosion as a factor in soil profile development. Nature 138:548.

Olson, C.G., and C.R. Hupp. 1986. Coincidence and spatial variability of geology, soils and vegetation, Mill Run watershed, Virginia. Earth Surf. Processes Landforms 11:619–629.

Penck, W. 1924. Morphological analysis of landforms. Translated by K.C. Boswell and H. Czech. Macmillan Co., London.

Pennock, D.J., and D.F. Acton. 1989. Hydrological and sedimentological influences on Boroll catenas, central Saskatchewan. Soil Sci. Soc. Am. J. 53:904–910.

Pennock, D.J., B.J. Zebarth, and E. deJong. 1987. Landform classification and soil distribution in hummocky terrain, Saskatchewan, Canada. Geoderma 40:297–315.

Reeve, R.C., and D. Kirkham. 1951. Soil anisotropy and some field methods for measuring permeability. Trans. Am. Geophys. Union 32:582–590.

Ruhe, R.V. 1956. Geomorphic surfaces and the nature of soils. Soil Sci. 82:441–455.

Ruhe, R.V. 1960. Elements of the soil landscape. Trans. Int. Congr. Soil Sci. 7th. 4:165–170.

Ruhe, R.V. 1975. Geomorphology. Houghton Mifflin, Boston.

Ruhe, R.V., and C.G. Olson. 1980. Soil welding. Soil Sci. 130:132–139.

Ruhe, R.V., and P.H. Walker. 1968. Hillslope models and soil formation: I. Open systems. Trans. Int. Cong. Soil Sci. 9th. 4:551–560.

Sinai, G., D. Zaslavsky, and P. Golany. 1981. The effect of soil surface curvature on moisture and yield — Beer Sheba observation. Soil Sci. 132:367–375.

Stone, J.R., J.W. Gilliam, D.K. Cassel, R.B. Daniels, L.A. Nelson, and H.J. Kleiss. 1985. Effect of erosion and landscape position on productivity of Piedmont soils. Soil Sci. Soc. Am. J. 49:987–991.

Troeh, F.R. 1964. Landform parameters correlated to soil drainage. Soil Sci. Soc. Am. Proc. 28:808–812.

Weyman, D.R. 1973. Measurements of the downslope flow of water in a soil. J. Hydrol. (Amsterdam) 20:276–288.

Whipkey, R.Z. 1965. Subsurface stormflow from forested slopes. Bull. Int. Assoc. Sci. Hydrol. 10:74–85.

Whipkey, R.Z. 1969. Storm runoff from forested catchments by subsurface routes. p. 773–779. *In* Floods and their computation, Vol. 2. Proc. Symp. Int. Assoc. Sci. Hydrol., Leningrad, USSR. August 1967. UNESCO/IASH, Belgium.

Whipkey, R.Z., and M.J. Kirkby. 1978. Flow within the soil. p. 121–144. *In* M.J. Kirkby (ed.) Hillslope hydrology. John Wiley & Sons.

Wilding, L.P., and L.R. Drees. 1983. Spatial variability and pedology. *In* L.P. Wilding et al. (ed.) Pedogenesis and soil taxonomy. I. Concepts and interactions. Elsevier, Amsterdam.

Wood, A. 1942. The development of hillside slopes. Geol. Assoc. Proc. 53:128–138.

Young, A. 1972. The soil catena: A systematic approach. p. 287–289. *In* W.P. Adams, and F.M. Helleiner (ed.) International geography. Int. Geogr. Congr., Toronto. Univ. Toronto Press, Canada.

Young, A. 1976. Tropical soils and soil survey. Cambridge Univ. Press, United Kingdom.

Zaslavsky, D., and A.S. Rogowski. 1969. Hydrologic and morphologic implications of anisotropy and infiltration in soil profile development. Soil Sci. Soc. Am. Proc. 33:594–599.

3 One Perspective on Spatial Variability in Geologic Mapping

H. W. Markewich

U.S. Geological Survey
Doraville, Georgia

S. C. Cooper

Blasland, Bouck & Lee Engineers
Syosset, New York

ABSTRACT

This paper discusses some of the differences between geologic mapping and soil mapping, and how the resultant maps are interpreted. The role of spatial variability in geologic mapping is addressed only indirectly because in geologic mapping there have been few attempts at quantification of spatial differences. This is largely because geologic maps deal with temporal as well as spatial variability and consider time, age, and origin, as well as composition and geometry. Both soil scientists and geologists use spatial variability to delineate mappable units; however, the classification systems from which these mappable units are defined differ greatly. Mappable soil units are derived from systematic, well-defined, highly structured sets of taxonomic criteria; whereas mappable geologic units are based on a more arbitrary heirarchy of categories that integrate many features without strict values or definitions. Soil taxonomy is a sorting tool used to reduce heterogeneity between soil units. Thus at the series level, soils in any one series are relatively homogeneous because their range of properties is small and well-defined. Soil maps show the distribution of soils on the land surface. Within a map area, soils, which are often less than 2 m thick, show a direct correlation to topography and to active surface processes as well as to parent material.

Rock units are classified based upon their stratigraphy. The basic geologic map unit, the formation, is identified by its lithologic characteristics and its stratigraphic position. A formation is commonly more heterogeneous than a soil map unit, even if the unit includes two or more soil series. Geologic mapping is a method of investigating the geologic history of an area and is based on lithologic assessments that include features indicating origin, relative age, and absolute age of mapped units. Unlike soils, a rock unit

---

Copyright © 1991 Soil Science Society of America, 677 S. Segoe Rd., Madison, WI 53711, USA. *Spatial Variabilities of Soils and Landforms.* SSSA Special Publication no. 28.

typically has no relation to processes now active at the earth's surface (excepting active floodplain, aeolian dunes, etc.). Lithologic descriptions of rock units include thickness, composition, bedforms and internal structures, fossil content, nature of contacts, and structures either defined by the unit (folds) or that cut the unit (faults, joints, etc.). The stratigraphic position of a rock unit in a sequence defines its age relative to superjacent and subjacent strata. The geometry of a unit gives information on its structural history. Because the origin of each rock unit and the geologic history of the map area can be inferred from a geologic map, the map can be viewed as a synthesis or summary of the geologic history of an area.

Although there are probably as many definitions of a geologic map or a soil map as the number of geologists or soil scientists asked, both types of maps serve as visual representations of either the rock units (parent material) or overlying soil that comprise the earth's surface in a specific area. The maps are the end-products of field observations and laboratory analyses of the respective scientists. The maps, however, differ in their respective functions. A geologic map summarizes the geologic history of an area — the length of time that is summarized is determined by the age of the rocks, years to billions of years. A soil map shows the distribution of soils on the landscape, by delineating specific soils, or groups of soils, that have developed on the underlying rock units and now form the thin outer layer or "skin" of the earth's crust. Soil age is limited not only by the age of the underlying parent material but by the length of time that the parent material has been subaerially exposed and by erosion rates. The age of most soils is less than 5 million years.

Common objectives of the field geologist and the soil scientist are to collect, interpret, and disseminate information concerning the types and spatial distributions of rock and soil that comprise the upper part of the earth's crust. The general scope of this work involves conducting detailed, step-by-step procedures to delineate the spatial variability of characteristic properties commonly used to classify rock or soil materials into formal, mappable units, and to construct a two-dimensional representation of their spatial distribution (a map). Units delineated on geologic maps and soil maps differ in scale (thickness, depth, etc.) and in history (origin, age, degree of consolidation, degree of deformation, etc.). In contrast to most rock units; most soils are less than 2 m thick, are not stacked in sequence, and have not been deformed (faulted or folded). In contrast with areas of extensive Quaternary deposits, erosional terrains generally have soils that are significantly younger (i.e., more than an order of magnitude younger) than the age of the area's rock units. For example, the Appalachian Piedmont soil residence times are about a million years, significantly less than the 300-million-year age of the parent material (Pavich, 1989).

A geologist is not only concerned with the rock units that are exposed at the land surface, but also with the geometry, origin, relative age, and absolute age of these units. By using measurements taken in the field, a geologist constructs a three-dimensional model of the map area, which is represented by a cross section that accompanies the geologic map. The cross section is the geologist's interpretation of the spatial distribution (vertical and horizontal) of rock units, which comprise the earth's crust in the map area, generally to depths of several hundred meters.

This chapter presents one perspective on how spatial variability is considered in geologic mapping; it does not address the problems of spatial variability in soil mapping. It reviews the formal stratigraphic classification system used by geologists to differentiate rock units, and discusses why spatial variability is only qualitatively considered in correlation of rock units. Descriptions are given for some of the steps taken in field mapping to identify and classify rock types, to delineate lithostratigraphic units (rock units), and to depict these units on a geologic map. Although some of the differences between geologic mapping and soil mapping are discussed, the discussion is not comprehensive. Rather, it is sufficient to provide a better understanding of the criteria used in differentiating rock units and geologic structures; and to emphasize differences in scale (between rock units and soils), in the range in ages of soils vs. the range in ages of rock units, and in the representation of time on the respective maps. A short discussion on derivative maps is also included.

CLASSIFICATION OF STRATIGRAPHIC UNITS

Unlike soils, rock units are generally part of a sequence of layers or strata that are bounded above and below by other rock units. Therefore, the classification system used for rock units is one based upon stratigraphy. A stratigraphic unit is defined as a naturally occurring body of rock or rock material distinguished from adjoining rock on the basis of some stated property or properties (North American Commission on Stratigraphic Nomenclature, 1983). As a result, rock strata can be classified and named on the basis of their lithology (lithostratigraphic unit), age (chronostratigraphic unit), fossil content (biostratigraphic unit), or other properties such as mineral content chemical composition, remnant magnetic properties, and so on. For most of these properties, there is no formal nomenclature.

The formal classification system and nomenclature currently used to classify stratigraphic units in geology are the products of about two centuries of gradual evolution. In the early 1930s, stratigraphers working in the petroleum industry recognized that rock units could not be forced into a formal classification scheme (Krumbein & Sloss, 1963). This led to the development of a dual classification scheme that separated rock units into distinct stratigraphic units. The basic criteria used to categorize strata required that; (i) formally recognized rock units be traceable or mappable over a distance, and (ii) strata be differentiated with respect to their positions in geologic time.

In 1941, Schenck and Muller published a paper that proposed two major categories of stratigraphic classification that incorporated both position in geologic time and physical characteristics that could be used to recognize a formal rock unit. The two categories were

1. "Rock-stratigraphic (lithostratigraphic) units" — mappable assemblages of strata that are distinguished and identified by objective physical criteria observed in the field and/or the subsurface.
2. "Time-stratigraphic (chronostratigraphic) units" — assemblages of strata that were deposited during distinct time intervals.

Following numerous revisions and changes, this first proposed dual classification system became what is now recognized as the North American Stratigraphic Code. The major objectives of the code are " . . . to provide a basis for systematic ordering of the time and space relations of rock bodies and to establish a time framework for the discussion of geologic history" (North American Commission on Stratigraphic Nomenclature, 1983).

Lithostratigraphic units are three-dimensional bodies of sedimentary, extrusive igneous, metasedimentary, or metavolcanic strata identified and delineated by observable lithologic characteristics and by stratigraphic position (North American Commission on Stratigraphic Nomenclature, 1983). Lithostratigraphic units generally conform to the Law of Surperposition, a principle of geology that states that in a sequence of strata that has not been overturned; the youngest strata are at the top and the oldest strata are at the base of the sequence. The lithostratigraphic unit serves as a basis for describing and interpreting local and regional structures, stratigraphy, economic resources, and geologic history.

Chronostratigraphic units are three-dimensional bodies of strata that are organized into units based on their age or their time of origin. The chronostratigraphic unit serves as a material reference for all rocks formed during a given time span (North American Commission on Stratigraphic Nomenclature, 1983). In theory, the boundaries of a chronostratigraphic unit are isochronous surfaces; in other words, these boundaries are theoretically independent of lithology, fossil content, or any other physical feature used to identify a stratigraphic unit. Thus the boundaries ideally represent everywhere the same horizon in time. In actuality, the geographic extension of these boundaries from a type section locale is influenced by interpretation of lithology, fossils, topography, and other physical features.

Table 3-1. Categories and ranks of formal stratigraphic units (modified from North American Commission on Stratigraphic Nomenclature, 1983)

Stratigraphic units		Geochronologic units
Lithostratigraphic (time of secondary significance)	Chronostratigraphic (relative time significant)	
Supergroup	Eonothem	Eon
Group	Erathem	Era
Formation†	System	Period
Member (or lens, or tongue)	Series	Epoch
Bed(s) (or flows)	Stage	Age

† Basic unit in geologic mapping.

Table 3-1 shows, in descending order the ranks of formal units that comprise the lithostratigraphic and chronostratigraphic categories of classification.

Lithostratigraphic Units

The formation is the primary unit in lithostratigraphic classification and geologic mapping (Table 3-1). A formation is a body of rock that is recognized by distinctive lithologic characteristics or by some degree of internal lithologic homogeneity; and that can be mapped or traced either from one outcrop to the next at the surface or from one borehole to the next in the subsurface. A formation can be characterized on the basis of one lithologic rock type, repetitions of two or more lithologic types, or extreme lithologic heterogeneity such as melange units that may in themselves constitute a homogeneous matrix of a variety of rock types when compared to adjacent rock units (North American Commission on Stratigraphic Nomenclature, 1983).

Distinctive lithologic characteristics used to define a formation include chemical and mineral composition, texture, sedimentary structures such as cross bedding, ripple marks, or mud cracks, and supplementary features such as color and fossil content. It should be noted that a formation is not recognized by its fossil assemblages alone. Distinctive electrical, radioactive, seismic, or other properties also may be useful indicators of lithology, but alone these properties do not adequately describe the lithologic character of a formation.

A formation has three dimensions including thickness, but thickness is not a determining factor used to define a formation. The thickness of a formation may range from hundreds of meters or more to a feather edge at the formation's margins. Additionally, subjacent or superjacent formations need not have the same general thickness. A formation that is 300 m thick may overlie another formation that is only 3 m thick.

A member is the next lower formal rank after formation, and it is recognized as a subdivision of a formation. Members are designated within a formation based on distinctive lithologic characteristics that enable the geologist to distinguish one member from adjacent parts of the formation. A member is established when it is useful to recognize and single out secondary units of special interest within a formation. Although members have only local significance in mapping, they must satisfy the same criteria of lithologic distinction and mappability used to delineate formations. However, the mappability of a member need not be at the same scale required for a formation, which is on the order of 1:25 000 (North American Commission on Stratigraphic Nomenclature, 1983).

A bed is the smallest formal unit recognized in lithostratigraphic classification. The designation of beds generally is limited to certain beds of economic importance (coal beds, oil- or water-bearing sand, ore horizons, etc.). Beds commonly are given names that have only local application, and are often excluded from the formal lithostratigraphic nomenclature.

Above the formation in rank is the group, which consists of two or more associated formations. Groups are identified for the purpose of demonstrating the natural relations of associated formations that have substantial lithologic features in common (North American Commission on Stratigraphic Nomenclature, 1983).

Last, a supergroup is a formal assemblage of related groups or of formations and groups. The rank of supergroup is used only where the designation serves a clear purpose.

Chronostratigraphic Units

Bodies of rock deposited during finite periods of time and preserved in the stratigraphic record are recognized as chronostratigraphic units. Two principle purposes served by chronostratigraphic classification are to (i) correlate rocks in one region with those in another based on their age equivalence (their contemporaneous origin), and (ii) place rocks of the earth's crust in a systematic, geochronologic sequence and indicate their relative position and age with respect to earth history as a whole.

The ranking of chronostratigraphic units is related to the time interval represented by the units as opposed to the thickness and/or areal extent of the strata comprising the unit. The fundamental chronostratigraphic unit is the system, which is comprised of rocks that represent a substantial time interval marked by a major episode in earth's history (Table 3-1). As a result, these rocks can serve as a worldwide chronostratigraphic reference unit.

Geologic maps do not generally depict chronostratigraphic units. Because they delineate time, these units are not always manifested physically. In a map area, the boundaries of lithostratigraphic units are commonly oblique to boundaries of chronostratigraphic units. In areas where there is abundant biostratigraphic information, where the lithostratigraphy is relatively straightforward or "layer-cake" (such as an undeformed Coastal Plain section or stacked basalt flows), and where there are numerous bounding unconformities, it is possible for a formation to be a lithostratigraphic, biostratigraphic, and chronostratigraphic unit. However, in most terrains, structural complexity, lack of adequate age data, and the distribution of outcrops result in lithostratigraphic units being selected as the basis for geologic mapping. Identification of chronostratigraphic units, therefore, results from biostratigraphic and lithostratigraphic studies augmented by other age-determining data.

CORRELATION OF STRATIGRAPHIC UNITS

Generally, a geologist does not directly consider the problem of spatial variability when mapping. A geologist is trained to picture stratigraphic units and structures in three dimensions, and to demonstrate spatial relations between distinct geologic features. In a sense, the concept of spatial variability is most useful in identifying and classifying, in vertical succession, different rock types as formal stratigraphic units. Characteristic features that can be used for identification include mineral composition, chemical composition, texture, structure, fossil content, geophysical properties, unconformable or cross-cutting relations, and age. Based on these criteria, a geologist can delineate individual stratigraphic units in vertical succession by recognizing surfaces or contacts between rock units that denote either periods of erosion, periods of nondeposition, or periods of gradual transition over an uninterrupted time interval.

The standard on which all stratigraphic units must be based in the "type section," which is an originally described sequence of strata that has been identified by its lithologic characteristics and not by the fossils contained within the strata (North American Commission on Stratigraphic Nomenclature, 1983). The type section is most commonly found in areas where the maximum thickness of the unit is shown, and where it is completely exposed. It serves as an objective standard by which spatially separated parts of a unit can be compared. The type section of lithostratigraphic units are never changed (North American Commission on Stratigraphic Nomenclature, 1983). Usually, exposures at other localities are selected as "reference sections." These become particularly useful in the event a type section is removed or altered.

Outside the area of the type section, a geologist identifies and maps rock types by their similar lithologic character and their stratigraphic position. The process of demonstrating the equivalence between spatially separated outcrops and the defined type section is referred to as correlation. Thus, stratigraphic correlation is used to show the spatial and temporal relations rather than variability between separated rock units. The North American Commission on Stratigraphic Nomenclature (1983) defines correlation as the demonstration of correspondence between two geologic units both in some defined property and in relative stratigraphic position.

Three specific types of correlation are lithocorrelation, biocorrelation, and chronocorrelation. Lithocorrelation demonstrates equivalence between units of similar lithology and stratigraphic position and is the most common correlation used in geologic mapping. Biocorrelation expresses equivalence of fossil content and biostratigraphic position. Chronocorrelation shows similarity in age and in chronostratigraphic position. Because of their use in geologic mapping, lithocorrelation and biocorrelation are discussed in the following two sections.

Lithocorrelation

Lithologic criteria such as color, texture, bedforms, etc., become less reliable as the distance between the unit and the type section increases. As distance between outcrops of map areas increases, so also does the importance of stratigraphic position as a correlation tool.

"Key beds" also become increasingly important with increases in the distance over which correlations are made. Stratigraphic horizons or key beds are generally used to delineate the upper and/or lower boundaries of the stratigraphic unit that is being mapped. Krumbein and Sloss (1963) categorized stratigraphic horizons or key beds into three classes, which include lithologic attributes of a section, biological (fossil) attributes, and structural discontinuities.

Lithologic key beds such as thin limestone beds, coal beds, thin layers of volcanic ash, and so forth, are used most often because they are relatively easy to observe and identify in outcrop, core, or on borehole geophysical logs.

Biologic key beds include interval zones and assemblage zones. Interval zones are defined as bodies of strata between two specified, documented lowest and/or highest occurrences of a single taxon. Assemblage zones are characterized by the association of three or more taxa, and can be represented by all kinds of fossils present in the strata or may be restricted to certain kinds of fossils (North American Commission on Stratigraphic Nomenclature, 1983).

The third type of stratigraphic horizon is structural discontinuities, which primarily include different types of unconformities. An unconformity is a substantial break or gap in the geologic record, where a rock unit is overlain by another rock unit that is not next in stratigraphic succession (Bates & Jackson, 1987). Unconformities are ideal for delineating the boundaries of a stratigraphic unit because they are sharp contacts that clearly show a change in the conditions that ended the formation of one rock unit and that began the formation of the subsequent rock unit. Structural discontinuities include features such as (i) an angular unconformity, which is an unconformity between two groups of rocks whose bedding planes are not parallel, or in which the older, underlying rocks dip at a different angle than the younger, overlying rocks; (ii) a disconformity, which is an unconformity where the bedding planes above and below the stratigraphic break are essentially parallel; and (iii) a nonconformity where stratified rock above the unconformity overlies unstratified igneous or metamorphic rock (Bates & Jackson, 1987). Not only can unconformities be used as a means of stratigraphic correlation but they also represent an interval of time, either of erosion or of nondeposition.

In addition to using lithologic characteristics, fossil content, stratigraphic position, and key beds as bases for correlation, a geologist may also use external characteristics such as topographic expression, or vegetation and soil coverage.

Topographic relief commonly reflects the original features and structures of stratigraphic units or reflects the resistance of a certain rock unit to erosion. For instance, in areas underlain by a resistant rock type, ridges or ledges will develop where the rocks crop out, whereas in areas underlain by less resistant rocks, valleys will develop. Alluvial terraces, glacial moraines, and sand dunes are examples of present topography mirroring the original form of the deposit.

Specific species or floral communities may preferentially grow on a certain lithology or a soil and be unique to a specific rock type or lithostratigraphic unit. Soils and vegetation become particularly useful in areas with limited outcrops or where the rock has been extensively weathered.

Biocorrelation

Although not everywhere present, fossils provide a good means of identifying units of similar age. Some species of fossil organisms span a considerable length of geologic time whereas others are more limited in the temporal distribution. A species that is characteristic of a specific geologic horizon and that occurs only in beds at the specific horizon are referred to as index fossils (Lahee, 1952). Index fossil species can be helpful not only in correlating separate exposed parts of the same bed in which the fossil is found, but also in correlating adjacent units above and below a bed of interest (Lahee, 1952). It should be noted, however, that fossil floras and faunas may vary laterally within a given bed or formation of the same age. Distinct faunal groupings or assemblages may even disappear so that at the proper stratigraphic position where the boundary between two units should occur, there may be a series of transitional faunas that are a mixture of faunas in both units.

COMMENTS ON CONSTRUCTION AND INTERPRETATION OF GEOLOGIC MAPS

Construction

A geologic map not only presents a description of rock units found in an area, but also describes the environment of deposition/formation during specific time intervals, and thus, serves as a basis for reconstructing the geologic history of the units.

Constructing a geologic map involves the compilation of all available outcrop and subsurface data from the area of study. An appropriate base map is either selected or drawn so that it is compatible with both the size and the scope of the study. The most appropriate base map is typically a topographic map of the area of interest. Control points that represent the locations of contacts or key beds are plotted on this base map. Information obtained in the field from borehole, well, or measured sections is summarized on the map at specific control points or in cross sections that accompany the map.

A geologic map produced for a study area may show solid or dashed lines that indicate the positions of contacts between formations. The attitude of a rock unit is represented on a map by a strike and dip symbol (illustrated on Fig. 3-1), which is critical to construction of the geologic cross section representing the geometry of the rock units in the subsurface. The dip is the angle of inclination of a bed or any other planar feature from a horizontal plane in the direction of the steepest descent. The strike, which is measured at right angles to the direction of the dip, is the intersection of the planar

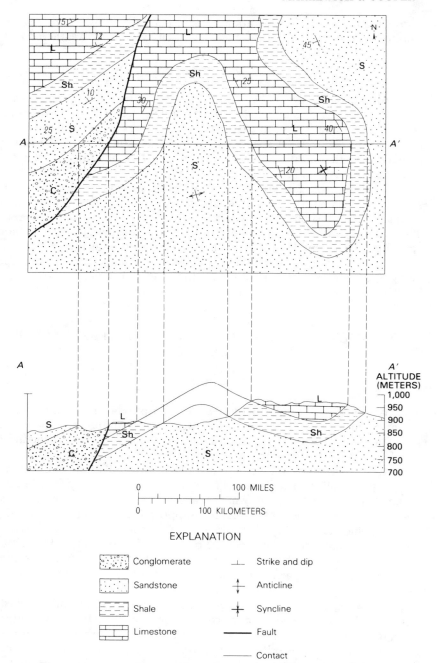

Fig. 3-1. Construction of a geologic cross section (A–A^1) from a geologic map that shows rock type, some strike and dip information, contacts, simple folds, and a fault. No age data are given. Stratigraphic units shown in cross section. C = conglomerate; S = sandstone; Sh = shale; and L = limestone.

feature with a horizontal plane. The longer line of the strike and dip symbol indicates the strike of the feature, and the shorter line indicates the direction of dip. The number shown next to the strike and dip symbol indicates the angle of the dip. Additionally, a geologic map will show the contacts of other important features such as intrusive bodies of rock, and the traces of faults in the study area (Fig. 3-1). An intrusion is a body of igneous rock that invades older rock, thus an intrusive contact between an igneous rock and some other rock indicates that the igneous rock is younger than the host rock.

Time is also an integral part of geologic maps. Geologic maps show not only the sequence of lithostratigraphic units (relative age), but also give whatever information is known about their actual age. Usually, the age of a lithostratigraphic unit is shown by a symbol that denotes the geologic period in which it was deposited; i.e., Jm might refer to a lithostratigraphic unit that has been identified as the Morrison Formation of Jurassic age. The Jurassic period, in turn, is only one of the periods that comprise the larger unit of time referred to as the Mesozoic era. If enough age information is available, it is sometimes possible to subdivide the periods and refine the age of the map units.

Missing time is also represented on geologic maps. Unconformities and disconformities represent periods of erosion or nondeposition, and are as important to construction of the geologic history as are the map units.

Included with all geologic maps is an explanation or legend, which usually is located at the side or at the bottom of a map. The explanation is used not only to define symbols shown on the map, but also to show the ages and relation among rock units. Typically, different rock units are shown on the map and defined in the explanation either as different colors or as different patterns. Rock units delineated on the map are nearly always arranged in the explanation in descending, chronological order from youngest to oldest regardless of the origin of the rocks. Thus, sedimentary, igneous, and metamorphic rock units that appear on a map are arranged chronologically to better facilitate interpretation of the structural and geologic history of the area.

Geologic cross sections are constructed through different parts of the study area after initial mapping in order to show the most logical interpretation of the geometry of the rock units in the subsurface. The horizontal scale is the same as the map scale; the vertical scale is exaggerated sufficiently to show the relation of time and rock units (Fig. 3-1). The line of intersection of the vertical plane and the land surface is referred to as the trace or line of section. Depths represented on geologic cross sections can be from a few meters to hundreds of meters. The cross section is constructed by projecting points of contact from the line of section onto a vertical plane. Field-measured values of the dips then are used to draw in the contacts between beds at their proper orientation along the line of section.

Interpretation

A geologic map displays information in a form that allows a geologist or anyone familiar with geology to conceptualize the spatial relations between rock units, to delineate structural features such as folds and faults, and to evaluate the origin of these geologic features. Examination of the areal distribution of structural features, rock units, and/or stratigraphic contacts on a geologic map and accompanying cross sections helps the geologist reconstruct the paleoenvironment and the structural history of the mapped area. Thus, a geologic map can be considered a summary of the geologic history of a given area.

The pattern of rock units and features on a geologic map along with other data (i.e., strike and dip of planar features, nature of contacts, lithology, fossil content, etc.) allow interpretation of specific events in the geologic history of the area and the chronological order in which the events happened. For instance, the user of the map can interpret from the map (i) the order in which strata were originally deposited, (ii) whether strata were folded and then faulted or vice versa, (iii) whether beds were laid down by a transgressive or regressive sea, (iv) the relative age of an intrusive rock unit, (v) whether uplift occurred before, during, or after erosion (or some combination of these), and so forth.

SOIL MAPS, GEOLOGIC MAPS, AND HOW THEY DIFFER

Soil maps show the distribution of soils in the landscape, and generally represent surface materials at depths of 5 m or less. The distribution of soils is largely in response to active surface and near-surface processes. The age (or residence time) of a soil depends upon many factors, one of which is its position on the land surface and how that position is affected by surface processes (erosion, dissolution, and/or sediment accumulation). In other words, the maximum age of a soil can be no greater than the age of the land surface upon which it has developed. The land surface may or may not approximate the age of the parent material.

Geologic maps differ from soil maps in that they represent, for a given area, the history of the earth's outer crust through time. No single characteristic or property can be used to define a body of rock as a formal geologic mapping unit, such as a formation. Likewise, no single measured value shown on a geologic map can accurately and completely define the geometric and compositional features of a lithostratigraphic unit.

Depending upon age, a lithostratigraphic unit depicted on a geologic map may or may not be related to geologic processes presently active in the map area. In general, lithostratigraphic units that are several tens of thousands to hundreds of millions of years in age have little relation to processes now active in the map area (e.g., deep sea carbonates that now form mountainous, glaciated terrain). In many areas, the rock units have undergone folding, faulting, and/or metamorphism. Although there are reported cases of soils being locally faulted, folded, and/or metamorphosed, it is not common, and in those cases, they are generally mapped as part of the associated lithostratigraphic unit.

A geologic map provides the basis for preparing second-order or derivative maps that can be used to illustrate the spatial variability of certain rock properties or features. Isopach maps show variability in thickness of a rock unit in the map area. Structure contour maps show the variation in altitude of the upper surface of a rock unit. Second-order maps can also be derived from compositional data. Examples include trace element maps and heavy mineral maps. Other second-order maps are generated to be used as tools in determining potential aquifers or for predicting water quality in a given rock unit. Maps for these purposes would include the variation in the ratio of coarse to fine fractions (sand and gravel to silt and clay) and/or the variation in phosphate content.

Another commonly generated second-order map is a landslide susceptibility map, which considers geologic, geomorphic, climatic, and biological factors. These factors include composition and structure of the parent material (i.e., mineralogy, strike and dip of beds and/or foliation, etc.), topography (slope), aspect, and vegetation. A landslide susceptibility map is particularly useful to agencies responsible for the construction and maintenance of roads and highways and for land-use planning. Economists and engineers commonly collaborate with the geologist in the assignment or ranking of the susceptibility units on these and other maps used by engineers and planners.

It is the second-order map that commonly addresses, in a more direct way, the problem of spatial variability within a rock unit. Generally, a geologic map based on lithostratigraphic units does not tell the user exactly what the parent material is at any specific location on the map. The user is only told that at any specific location, the properties of the material will be within the range of properties as described for that unit.

SUMMARY STATEMENT

Pedogenesis can be defined as the chemical and physical alteration of material at and near the earth's surface. A soil map depicts the variation in soils on the land surface as a function of the five soil-forming factors (i) climate, (ii) topography, (iii) organisms, (iv) parent material, and (v) time (Jenny, 1941). A geologic map depicts not only the distribution and composition of the parent material, but, by delineation of lithostratigraphic units, it shows their origin, structural history, and age. In defining lithostratigraphic units, spatial variability is considered for all components of the unit — composition, structure, degree of metamorphism, stratigraphic position, etc. — but generally no attempt is made to quantify the variability. A geologic map is a synthesis of the temporal as well as the spatial variability of rock units in the map area. It is, in a very real sense, a highly condensed book that relates the geologic history of a specific area.

REFERENCES

Bates, R.L., and J.A. Jackson (ed.). 1987. Glossary of geology, 3rd ed. Am. Geol. Inst., Alexandria, VA.

Jenny, H. 1941. Factors of soil formation: A system of quantitative pedology. McGraw-Hill Book, Co., New York.

Krumbein, W.C., and L.L. Sloss. 1963. Stratigraphy and sedimentation, 2nd ed. W.H. Freeman and Co., San Francisco, CA.

Lahee, F.H. 1952. Field geology, 5th ed. McGraw-Hill Book Co., New York.

North American Commission on Stratigraphic Nomenclature. 1983. North American stratigraphic code. Am. Assoc. Petrol. Geol. Bull. 67:841–875.

Pavich, M.J. 1989. Regolith residence time and the concept of surface age of the Piedmont "Peneplain." Geomorphology 2:181–196.

Schenk, H.G., and S.W. Muller. 1941. Stratigraphic terminology. Geol. Soc. Am. Bull. 52:1419–1426.

4 Scientific Methodology of the National Cooperative Soil Survey

Stephen L. Hartung
University of Nebraska
Lincoln, Nebraska

Steven A. Scheinost
USDA-SCS
Wahoo, Nebraska

Robert J. Ahrens
USDA-SCS
Lincoln, Nebraska

ABSTRACT

Reliable soil surveys can be made at reasonable cost because the location of soils are predictable on the landscape. This soil-landscape relationship is the scientific basis that makes it possible to produce a soil mapping model. The soil scientist designs map units based on these models. Saunders County, NE, is used to illustrate the application, design, and redesign of map unit models. Data from pedon descriptions, transects, field notes, and laboratory analyses and knowledge of landscape patterns, geology, and climate are used to develop the models. The model for each map unit is tested during mapping and is adjusted as needed. Well-conceived and tested map units derived from models result in reliable and accurate soil surveys.

Scientific methodology uses observed and/or inferred data to develop a model or hypothesis about the nature or principles of the object being investigated. This model is tested by observing or studying its predictions. If the predictions are correct the model is proven. If the model is disproven then a new hypothesis or a revision of the former one must be formulated and evaluated.

Soils do not occur randomly and thus have a degree of predictability. All soil properties cannot be accurately predicted, but where the five soil-forming factors (climate, living organisms, parent material, time, and relief) are similar, similar soils should be formed. The cumulative but differing

---

Copyright © 1991 Soil Science Society of America, 677 S. Segoe Rd., Madison, WI 53711, USA. *Spatial Variabilities of Soils and Landforms.* SSSA Special Publication no. 28.

effects of these factors on soil formation are commonly expressed as observable properties. Because of the observable relationships of soil properties we say that soils with discrete sets of properties have a degree of predictability on the landscape (Miller et al., 1979; Soil Survey Staff, 1980a,b; Witty & Arnold, 1987). This is the scientific basis of the soil survey.

Field soil scientists use this methodology almost daily when mapping soils. However, most are not consciously aware of the more formal aspects of the methodology. Field soil scientists are keenly aware of the predictability of soils on the landscape. While this awareness is the expression of the soil survey's scientific basis, the reliability of the predictions obtained is a function of the soil scientists' experience, knowledge, abilities, and the complexity of the mapping area (Miller et al., 1979). In order to successfully produce a usable soil map, the soil scientist must be able to consistently interpret and predict the relationship between soils and landscape. The soil scientist is not physically able to probe every acre in the survey area, or to use random sampling techniques that allow each member of the soil population equal probability of being sampled. By using soil–landscape predictability, time and money are saved because soil boundaries are predicted and checked by sampling to either confirm or deny the map unit model and soil line placement.

The map unit model is an expression of the soil scientist's hypothesis of soil genesis on the landscape. This model is an attempt to accurately represent the natural soils on the landscape and to balance and reconcile various cultural, photogrammetric, and taxonomic properties of these soils. The model is modified by the intended uses and interpretations that best meet the needs of various users of the soil survey. It is also modified by the scale at which the soils will be mapped. Legibility of maps and utility of the map legend to users causes and may encourage the soil scientist to include areas of dissimilar soils in the map unit (Cline, 1974a,b; Miller & Nichols, 1979). These areas of dissimilar, as well as similar, soils, and how we categorize them, are a reflection of both our taxonomic system and the interpretations we make. The taxonomic system has a profound affect on how we look at soils and is a modifying influence on map unit design and control.

MAP UNIT DESIGN MODEL

The decision to remap, update and/or recorrelate the 1965 soil survey of Saunders County, NE (Elder et al., 1965), provided a unique opportunity and area to apply soil survey models. This soil survey lacked useful maps and interpretive data to make some needed evaluations of soil resources. Field mapping was completed over a period of 20 yr (1939–1959) and was finished before *Soil Taxonomy*. The original identification legend contained over 600 map units, many of which were poorly defined. These 600 map units were reduced to 88 by combination during the final correlation. Some soil areas were examined in detail and others were not. A refinement of the map unit model was needed.

Saunders County (Fig. 4–1) has average annual precipitation of 69 cm/yr. The soil temperature regime is mesic and the soil moisture regime is typically udic.

Two large associations from the general soil map in the 1965 survey of Saunders County (Elder et al., 1965) were selected to begin investigation of the previous map unit models. These were the Sharpsburg and the Sharpsburg-Fillmore Associations (Elder et al., 1965).

The Sharpsburg Association is on loess-covered uplands. The loess overlays older reddish loess and/or glacial till at depths of 1.5 to more than 4.5 m. The uplands are mostly gently sloping to moderately steep with rapid runoff during heavy rains. Erosion is a major problem (Elder et al., 1965).

The Sharpsburg-Fillmore Association is in the Todd Valley, a loess-filled paleovalley of the Platte River. The loess is underlain by alluvial or wind-reworked sand at depths ranging from 2 to 6 m. The valley is nearly level and has many slight knolls and depressions in closed drainage basins. Moisture not entering the soils as rain collects in the depressions (Elder et al., 1965).

The predominant soil series mapped in both associations was Sharpsburg (fine, montmorillonitic, mesic Typic Argiudolls). This soil is moderately well drained and has a mollic epipedon 25 to 60 cm thick. The Fillmore Series (fine, montmorillonitic, mesic Typic Argialbolls) is poorly or very poorly drained. This soil is in depressions in the Todd Valley. In the Sharpsburg-Fillmore Association, Sharpsburg soils were mapped primarily on 0 to 2% slopes. In the Sharpsburg Association on the uplands, Sharpsburg soils were mapped primarily on 4 to 6 and 6 to 12% slopes (Elder et al., 1965). Three map units from these two associations were selected for preliminary study. In the Todd Valley, Sharpsburg silty clay loam, 0 to 2% slopes (ShA), was selected (Elder et al., 1965). In the uplands, 6 to 12% slopes, eroded (ShD2),

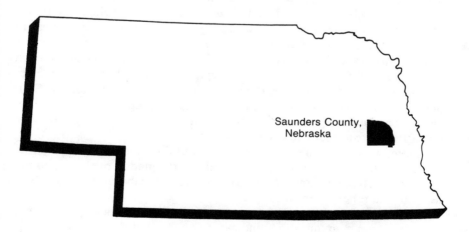

Fig. 4–1. Location of Saunders County, NE.

and 6 to 12% slopes, severely eroded (ShD3), were selected (Elder et al., 1965). Because of questionable definition and lack of adequate interpretations and transect data, these 1965 map units models required testing in the field by transecting.

TESTING OF MODELS AND TRANSECTING

The basic concepts of the soil-landscape relationships expressed in map units from the 1965 survey were identified in the hypotheses to be tested. Each map unit was evaluated both in the office and field. The soil survey party members used their experience and knowledge of soil-landscape relationships gained from mapping soils in similar situations in adjoining counties to select areas from which randomly selected transects were produced. Aerial photography and topographic maps were also used to help select these areas. The boundaries of these areas were physically measured and identified on the land surface. Ten-point transects were run from boundary to boundary, perpendicular to the axis of the map unit. The distance between points on the transects ranged from 14 to 33 m, but the average was 20 m. Abbreviated pedon descriptions were prepared for soils at each point on the transect. The most commonly observed characteristics were slope, horizonation, moist colors, depth, color of mottles, and texture.

The pedon at each site on the transect was tentatively classified in the field and a "typical" pedon from each transect was selected. A complete description (SCS SOI-232) of this pedon was then written in the field.

Each transect was analyzed in the office for percentage of pedons that classified the same as Sharpsburg, pedons that classified differently, and whether the slope was within the range definitive for the map unit. Five to ten transects were taken from each map unit and, after averaging, a "typical" pedon from a typical transect was selected for sampling and laboratory analyses. These pedons were then sampled in the field with the assistance of Soil Conservation Service National Soil Survey Laboratory personnel. Each site was excavated with a backhoe to a depth of 1.8 m and probed to greater depths using a Giddings probe (Giddings Machine Co., Fort Collins, CO). Samples were taken for complete characterization studies from the typical pedon. Satellite samples for determining the range in characteristics were taken from areas near each typical pedon.

After analysis of the transects and pedons (Hartung et al., 1988-1989, unpublished data), it was possible to evaluate the models of the previously identified map units from the 1965 soil survey. The laboratory data will be used to fine tune the new models.

Analysis of data from five transects on the Sharpsburg silty clay loam, 0 to 2% slopes map unit (ShA) indicated only 26% of the soils were in the same taxonomic class as Sharpsburg. The old map unit model was based on the soils being either Sharpsburg or very similar soils. The remainder of the soils were classified as: 46% Aquic Argiudolls, 17% Mollic Hapludalfs, and 11% Typic Argialbolls (Fig. 4-2). They were all in the fine, montmorillonitic, mesic family. "Soil Taxonomy" is not always the best method for testing map-unit models, but in this case the taxonomic differences also represented interpretation differences. The redesigned map unit models will be based on the new data.

Soils tentatively classified as Aquic Argiudolls had mollic epipedons with an average thickness of 90 cm; and mottles within 50 cm of the surface or directly below the mollic epipedon (indicative of an aquic subgroup). Those tentatively classified as Typic Argiudolls had mollic epipedons with an average thickness of 38 cm with a range of 25 to 60 cm; and had mottles with a range of depths of 60 to 100 cm, but were more than 15 cm below the mollic epipedon. The Aquic and Typic Argiudolls were on similar landscapes. The soils classified as Mollic Hapludalfs were thought to be eroded mollisols. They had an average of 15 cm of mollic color, and were tentatively classified according to the diagnostic horizons present (Lewis & Witte, 1980; Soil Survey Staff, 1975, 1980a, 1987), which were an ochric epipedon and an argillic horizon.

The presence of soils with overly thickened mollic epipedons, ochric epipedons, mottles diagnostic of aquic subgroups, and the interpretation differences associated with these morphological properties has made it necessary to further consider refining the previous map unit model. Piezometers will be installed to evaluate the moisture conditions of these soils.

Slope data indicated 63% of the unit was within the slope range (0-2%) and 37% had greater slope gradients with a high range of 8%.

Data for Sharpsburg silty clay loam, 6 to 12% slopes, eroded (ShD2); and Sharpsburg silty clay loam, 6 to 12%, severely eroded (ShD3) indicated similar compositions. Data were from seven and five transects respectively.

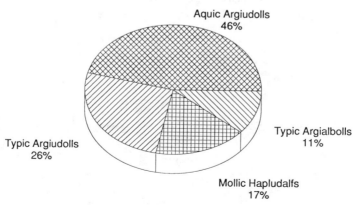

Fig. 4-2. Map unit composition: Sharpsburg sicl, 0 to 2% slopes.

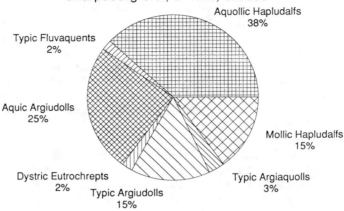

Fig. 4-3. Map unit composition: Sharpsburg sicl, 6 to 12%, eroded.

Pedons were tentatively classified. The ShD2 map unit had only 15% classify in the same taxonomic class as Sharpsburg. The remainder were 38% Aquollic Hapludalfs, 15% Mollic Hapludalfs, 25% Aquic Argiudolls, and 7% other soils (Typic Argiaquolls, Dystric Eutrochrepts, and Typic Fluvaquents) (Fig. 4-3). The ShD3 map unit had only 16% Typic Argiudolls; but had 30% Aquollic Hapludalfs, 30% Mollic Hapludalfs, 22% Aquic Argiudolls, and 2% Typic Argiaquolls (Fig. 4-4). The soils from these map units were all in the fine, montmorillonitic, mesic family except the Dystric Eutrochrepts and Typic Fluvaquents, which were in the fine-silty, mixed, mesic family.

The taxonomic distinctions shown in the ShD2 and ShD3 map units also represented interpretation differences and will have to be addressed in redesigned map unit models.

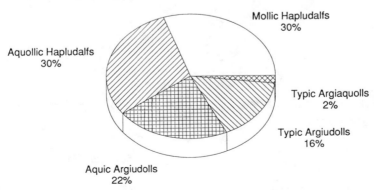

Fig. 4-4. Map unit composition: Sharpsburg sicl, 6 to 12%, severely eroded.

Aquollic Hapludalfs and Mollic Hapludalfs from these units averaged 15 cm of mollic colors. The Aquollic Hapludalfs had properties diagnostic of an aquic subgroup, mottles with chroma of 2 or less in the upper 25 cm of the argillic horizon. The Mollic Hapludalfs had mottles with chroma of 2 or less at depths of 43 cm or more below the soil surface (26–35 cm below the upper boundary of the argillic). The Aquollic and Mollic Hapludalfs were on similar parts of the landscape. They appeared to differ only in the depth to mottles and were hypothesized to be similar and to have similar moisture conditions. These soils were considered to be Aquollic Hapludalfs pending data from piezometers that will be installed to determine differences in moisture conditions.

The Aquic and Typic Argiudolls in these map units appeared to be similar, except in depth of mottles. Both had mollic epipedons with an average thickness of 60 cm, were on similar landscapes, and were considered to be Aquic Argiudolls pending piezometer data on moisture condition differences.

Close similarities between these two map units have made it necessary to consider refining the 1965 map unit models.

Slope data for ShD2 indicated that 73% of the unit was within the 6 to 12% slope range, 26% in the less-sloping range, and 1% in the more-sloping range. Seventy percent of the less-sloping areas were in the 5 to 6% slope range. The ShD3 had 54% of the unit within the slope range and 46% in the less-sloping range. Twenty-five percent of the less-sloping soil was in the 5 to 6% slope range. Between the two units only three transect points had slopes of 11% or greater.

REDESIGN OF MAP UNIT MODELS

The needs, uses, and objectives of the new soil survey of Saunders County were identified by an interdisciplinary group from the Soil Conservation Service, University of Nebraska-Lincoln, and the Lower Platte North Natural Resources District. These three entities along with the Lower Platte South Natural Resources District and the Saunders County Board of Supervisors were the pre-eminent cooperating agencies. A public meeting was held to get other user and interested-party comments about the soil survey. A questionnaire was also sent to users and potential users. People who were contacted and returned comments included county and state tax assessors, bankers, farm managers, real estate appraisers, and farmers. Using this and other information, the soil survey party determined the amount of mapping detail and interpretation refinement needed to adequately and accurately represent soil–landscape relationships and meet the objectives of the survey.

The soil survey party and leader normally have a number of options for designing and naming map units. The type of map unit is one of these. Generally, map units with more than one taxa are considered less refined than those identified by a single taxon (Soil Survey Staff, 1980b). The level of soil taxa used to name the map unit is another item that may be varied. Soil series are more narrowly defined than higher categories and provide more precise information and interpretations. Our transect data and data from other sources (Cline, 1977a) indicate complexes rather than consociations may better represent the soil patterns in some areas. In this county, soil series phases or complexes of series were determined to best fit the objectives of the survey.

The map scale and intensity of field investigations determines or helps determine the purity of the map unit. The soil scientist should produce a legible map uncluttered by unnecessary, small and confusing delineations. This is done by balancing map unit design, map scale, and intensity of field procedures to natural soil patterns, user needs, and standards of purity.

Redefined map unit models are proposed for Saunders County. The models will accommodate increased knowledge of soil-landscape relationships, a necessary interpretation refinement for the soil survey, a minimum delineation size of 1.2 to 2.0 ha, and the 1 to 12 000 map scale.

The redesigned map unit model of ShA (1965 survey) will still have slopes of 0 to 2%. Nearly all of the soils with ochric epipedons were within the 37% part of the 1965 map unit with greater than 2% slopes. These areas will be treated as small slope inclusions or delineated as part of a 2 to 5% slope map unit depending on area size. The name of the major component of the 0 to 2% map unit is not yet identified. After we have evaluated the water states of the soils with mottles within 50 cm of the surface we will be able to decide if a consociation or complex is appropriate.

The ShD2 and ShD3 map units will be combined into one unit with 5 to 11% slopes. Transect data indicated these slopes were most representative of the units and fit the needs and objectives of the new survey. Areas of less slope will be included or separated and placed in a 2- to 5%-slope map unit. These units were dominated by Aquollic-Mollic Hapludalfs with 10 to 17 cm of mollic colors; but a significant portion were Aquic-Typic Argiudolls with mollic epipedons with an average thickness of about 60 cm. The new map unit model proposed for testing will be a complex of a soil series representing Aquollic or Mollic Hapludalfs and a soil series representing Aquic or Typic Argiudolls. This map unit will have inclusions of soils with mottles higher or lower in the profile depending on the subgroup used. Again the Aquic and Aquollic subgroups need further testing and results may alter the map unit model. Major interpretation differences between the two components with mollic and ochric epipedons should include yield distinctions and differences in herbicide/fertilizer application based on their differing organic matter content (Nebraska Cooperative Extension Service, 1989).

The redesigned map unit models are being tested in the field with a number of aids to facilitate mapping. The new aerial photos, at a scale of 1 to 12 000, show more detail than the 1965 maps with scales of 1 to 20 000. The 1965 mapping is used as an indicator of slope breaks or major parent material-soil-landscape changes. Another aid is topographic maps that have been enlarged to the same scale as the aerial photos.

After the map unit models have been tested in the field by mapping, additional transects will provide further refinement. These transects will be used to compare differences between map unit models from the 1965 and the new survey. Data obtained in this manner should help determine if the newly designed models are an improvement over the older concepts.

The transects can also be used to estimate map unit composition (Arnold, 1977; Johnson, 1961). Transect data should enable us to make statements about the map unit composition using probabilities and percentages. For instance, transect date from the ShD2-ShD3 combined unit might allow the statement that at the 90% probability level, Aquollic (Mollic) Hapludalfs are 58 ± 3% of the map unit. It is hoped this will greatly facilitate writing map unit descriptions and increase reliability.

MAP UNIT DESCRIPTIONS

Map unit descriptions include information from typifying pedons, field notes, transects, and any other pertinent data that help describe the nature of the soils and landscape characteristics of the map unit. The map unit description includes characteristics of the dominant pedon(s) and other soils occurring within the map unit. Map units describe distinctive features important to crop management and other management considerations. Landscape position or distinctive patterns of map-unit components and map-unit to map-unit landscape relationships help distinguish characteristics used to separate map units and are an essential part of the map-unit description.

PROGRESSIVE CORRELATION

Progressive correlation is an integral part of the methodology of the National Cooperative Soil Survey. Each time a soil scientist examines a soil, he/she is testing and refining the soil-map unit model and practicing progressive correlation. Progressive correlation means that during each field review the taxonomic and map units recognized since the last review are evaluated and approved. Soil survey interpretations are developed and updated, soil investigations are completed; and the soil survey manuscript is developed concurrently with mapping. Beginning with the initial field review and lasting until the final correlation, soils staffs from state to national levels are present to maintain quality control and assure that the soil survey meets standards of the National Cooperative Soil Survey.

OTHER DATA

The fact that Saunders County, NE, is located near the National Soil Survey Center; and the University of Nebraska allows for additional soils research. Some of the research projects include: WEPP (Water Erosion Prediction Program), fertility classification systems (e.g., FCC-Soil Fertility Capability Classification System), and landcover and soil-chemistry relationships studies. This research should enhance the data base for the county and for soils in general, and may result in further refining the models.

SUMMARY

The scientific basis of the soil survey is the predictability of the location of soils on the landscape. This relationship makes it possible to produce a soil mapping model that is tested in the field during mapping activities. Testing allows refinement of the model. The model cannot be reliable without pedon descriptions, transects, field notes, and a knowledge of the geomorphology of the area. Map units conceived and tested using well-conceived models with a scientific base result in reliable and accurate soil surveys.

REFERENCES

Arnold, R.W. 1977. CBA-ABC: Clean brush approach achieves better concepts. p. 61–91. *In* Proc. New York Conf. Soil Mapping Quality Procedures, Bergmanin East, NY. 5–7 Dec. 1977. Cornell Univ., Ithaca, NY.

Cline, M.G. 1977a. The soils we classify and the soils we map. p. 5–20. *In* Proc. New York Conf. Soil Mapping Quality Procedures, Bergmanin East, NY. 5–7 Dec. 1977. Cornell Univ., Ithaca, NY.

Cline, M.G. 1977b. Thoughts about appraising the utility of soil maps. p. 251–273. *In* Soils resource inventories, Mimeo no. 77–23. Cornell Univ., Agron. Dep., Ithaca, NY.

Elder, J.A., T.F. Beesley, and W.E. McKinzie. 1965. Soil survey of Saunders County, Nebraska. Univ. of Nebraska, Conservation and Survey Div., and USDA-SCS, U.S. Gov. Print. Office, Washington, DC.

Johnson, W.J. 1961. Transect methods for determination of the composition of soil mapping units. USDA-SCS Soil Survey Technical Notes. U.S. Gov. Print. Office, Washington, DC.

Lewis, D.T., and D.A. Witte. 1980. Properties and classification of an eroded soil in southeastern Nebraska. Soil Sci. Soc. Am. J. 44:583–586.

Miller, F.P., D.E. McCormack, and J.R. Talbot. 1979. Soil surveys: Review of data collection methodologies, confidence limits, and uses. p. 57–65. *In* F. Zwanzig (ed.) The mechanics of track support, piles, and geotechnical data. Transportation Research Record 733. TRB-NAS-NRC, Washington, DC.

Miller, F.T., and J.D. Nichols. 1979. Soils data. p. 67–89. *In* M.T. Beatty et al. (ed.) Planning the uses and management of land. Agronomy Monogr. 21. ASA, CSSA, SSSA, Madison, WI.

Nebraska Cooperative Extension Service. 1989. A 1989 guide for herbicide use in Nebraska. Nebraska Ext. Circular 89–130.

Soil Survey Staff. 1975. Soil taxonomy: A basic system of soil classification for making and interpreting soil surveys. USDA-SCS Agric. Handb. 436. U.S. Gov. Print. Office, Washington, DC.

Soil Survey Staff. 1980a. Soil survey manual. USDA-SCS Handb. 18. U.S. Gov. Print. Office, Washington, DC.

Soil Survey Staff. 1980b. USDA-SCS. National soils handb. 430–VI. U.S. Gov. Print. Office, Washington, DC.

Soil Survey Staff. 1987. Keys to soil taxonomy. USDA-SMSS-AID. Techn. Monogr. no. 6. Cornell Univ., Ithaca, NY.

Witty, J.E., and R.W. Arnold. 1987. Soil taxonomy: An overview. Outlook Agric. 16:8–13.

5 Statistical Procedures for Specific Objectives

Dan R. Upchurch

USDA-ARS
Lubbock, Texas

William J. Edmonds

Virginia Polytechnic Institute and State University
Blacksburg, Virginia

ABSTRACT

The field of statistics is devoted to understanding variability in populations and then using this understanding to compare populations. The objective of this chapter is to give a brief overview of some of the statistical procedures that are used in soil science to quantify differences between soil types and to describe the spatial and temporal variability that is present within mapping units. Procedures that are described include a group of tools called parametric and nonparametric statistics. These all require a prior knowledge of, or an assumption about the probability distribution of the population. These tools also require that the samples be temporally and spatially independent. The second group of tools described is called geostatistics. Geostatistics are based on the theory of regionalized variables that combines our understanding of the continuous nature of geologic properties in space with the random variation that is present in spatially separated samples. Procedures included in geostatistics are extensions of the classical statistical tools with the assumption of sample independence removed. A discussion of the problems and procedures associated with sampling for spatially variable soil properties precedes the description of the various statistical procedures. This chapter provides a direction for further study of the literature by those concerned with this subject.

The field of statistics is devoted to understanding variability in populations and using this understanding to compare populations. Variability has been partitioned into two broad classes, random and systematic, with the division based on the source of the error that produces the variation. Systematic variability is that variability that can be attributed to a known cause, understood, and predicted. For example, soil properties are known to vary

---

Copyright © 1991 Soil Science Society of America, 677 S. Segoe Rd., Madison, WI 53711, USA. *Spatial Variabilities of Soils and Landforms.* SSSA Special Publication no. 28.

systematically as a function of topography, vegetation, climate, and parent material. When variability cannot be related to a given cause, it is called chance or random variability. In general, natural or undisturbed soils would be expected to have a larger systematic than random variability. However, this relationship between systematic and random variation will likely be scale dependent.

In classical statistical analysis, it is the random variability that has received the most attention. It is assumed that the systematic variability can be extracted with appropriate experimental design or will be constant across the treatments. Thus it would cancel when comparisons are made. Assuming that systematic variability has been controlled, the resulting variability in measurements would be considered random. Replication of treatments is intended to force this assumption to be true. A well-defined group of statistical procedures exists for analysis of results when the variability can be attributed to random processes and other assumptions are satisfied. It is important to note that although there are only a few assumptions required, they are not often fully achieved. This is particularly true for the assumption of normality.

Selection of appropriate procedures for describing and comparing soil populations is contingent upon attributes of the sample. The use of parametric estimators and tests assumes normality, randomness, equal variance, and independence of observations. Deviation from the first three assumptions results in approximate inference. When these assumptions are not valid and data transformations fail to produce appropriate results, nonparametric or distribution-free procedures are valid alternatives. Nonparametric procedures have a broader range of applicability because they require fewer assumptions. Some require only that populations be continuously distributed. Nonparametric procedures are less powerful when assumptions of normality, randomness, independence, and equal variance are valid. Conversely, nonparametric procedures are more powerful to the degree that these assumptions are violated.

Tests of normality and equal variance are computed by several statistical programs; e.g., SAS Institute, Inc. (1985a,b), which is available for both mainframe and microcomputers and SYSTAT, Inc. (Wilkinson, 1988) which runs on a microcomputer.

A test statistic and probability level for normality are computed by the normal option of the UNIVARIATE procedure (SAS Institute, Inc., 1985a). The Kolomogorov-Smirnov test statistic and probability level are computed by the NPAR procedure (Wilkinson, 1988). It is important to note that the central limit theorem allows the use of procedures requiring normality for describing nonnormal populations if large samples are available (Lentner, 1984). Examples of the use of the Shapiro-Wilks test for normality of selected soil properties are given by Edmonds et al. (1988), Edmonds and Lentner (1987), and Wright and Wilson (1979).

A test of equal variance can be made using $F = s_1^2/s_2^2$ with $n - 1$ degrees of freedom for Population 1 and $m - 1$ degrees of freedom for Population 2 where s_1^2 and s_2^2 are estimated variances for Populations 1 and

2, respectively. Tables of percentiles of the F distribution are given by Lentner (1984, p. 342), Ott (1984, p. 700) and Freund and Smith (1986, p. 513). The F' statistic and probability level computed by the TTEST procedure (SAS Institute, Inc., 1985b) are for $F = s_1^2/s_2^2$. Equal variances among several populations can be tested by the F_{max}-test and Bartlett's test (Sokal & Rohlf, 1969, p. 371).

The role of randomness in statistical inference is discussed by Lentner (1984, p. 10), "... there is always a possibility of drawing erroneous inference from any sample. Only with random sampling can we validly apply the rules of probability to calculate the chance of such errors. Knowing the error rates is essential if a meaningful "degree of confidence" is to be attached to statistical inference."

Nominal, ordinal, interval, and ratio scales are used in soil surveys. Nominal data are observations recorded according to name. They may be numbers that serve only as names. Ordinal data refer to ordered or ranked observations. Ordinal data may consist of numbers, but these numbers have no meaning beyond rank. Interval and ratio data are observations recorded according to scales that have all the properties of the ordinal scale and have numerical measurements of a quantitative nature. The interval scale lacks a true zero point. Parametric procedures must be based on data recorded according to interval or ratio scales. Data recorded according to nominal and ordinal scales must be analyzed by nonparametric procedures. However, specific nonparametric procedures are valid for data recorded according to any scale (Lentner, 1984).

One of the assumptions made in the previous discussion dealt with sample independence. When the observed variation in a sample is spatially correlated the assumption of sample independence is not valid, even if random sampling is used. For example, if the difference between measurements made at locations a specific distance (or direction) apart is a function of that distance (or direction), they cannot be considered independent measurements. Measurements that are correlated temporally would result in the same violation of the assumption of independent measurements. Again, there is a group of statistical tools that are applicable to the analysis of spatially dependent data. Part of this group of tools has been referred to collectively as geostatistics, which includes the semivariogram, kriging, covariogram, cokriging, and block kriging. These procedures are intended to ultimately lead to interpolation between observed data points. They may also be used to provide estimates of population means and variances. Another group of procedures that are applicable to spatially or temporally correlated data are those referred to as time series analysis, including autocorrelation, spectral density function, and cross-correlation function. These procedures are intended for the analysis of sequences of data. In general, they are used to determine if the sequence is random or possesses a trend. When a trend is present, they are used to determine the nature of the trend; e.g., is the trend cyclic?

This chapter is intended to be a general discussion of the broad range of statistical procedures available for the analysis of data obtained from soil populations. Our desire is not to provide a complete description of all of the procedures we will mention or to develop an exhaustive listing of all possible procedures, but rather to introduce those procedures we consider appropriate. The reader is referred to more in-depth treatments of each of the procedures to obtain a complete understanding of their application.

SAMPLES AND SAMPLING

Any discussion of statistical-inference building must begin with a discussion of sampling. Prior to analysis, a set of data must be obtained and the procedures employed in obtaining the data must be carefully planned. The procedures employed in sample collection often establish the basis for accepting or rejecting the use of a specific analysis procedure. Sampling procedures can support or invalidate the assumptions required for a particular analysis tool to be used appropriately.

There are three distinct issues involved in sampling a soil for a particular property: (i) location of the point to be sampled, (ii) size (volume) of individual samples and (iii) number of samples (observations) to be collected. Any sampling design must consider the resources available for sampling, the requirements of the project, characteristics of the property to be measured, and the intended statistical analysis to be performed on the results. Sampling design will often be a compromise, influenced by each of these needs. The expertise of both the statistician and that of the investigator should come to bear in developing a sampling design.

The effect of various sampling approaches can be seen by overlaying several sampling procedures on an area with a known spatial distribution of property values. For example consider a field represented by Fig. 5-1 in which some property of the soil varies in the manner described by the isolines. Not all sampling schemes will detect the same degree of variation. It should be recognized that the following examples are contrived to illustrate a point. However, the outcomes described are real possibilities when a similar situation is encountered. In Fig. 5-1 the property is stable in the upper right-hand corner of the field. Near the lower left of the field there is a rather gradual change in the property in contrast to the lower right where the value changes rapidly. In the upper left, there is a local extreme surrounded by a fairly stable area.

Sampling schemes can be split into two categories, random and systematic. Traditionally, statisticians have preferred a random sampling scheme for locating points to be sampled because the assumption of sample independence was to be made in the analysis procedure. However, the fact that the sample locations are randomly chosen does not guarantee sample independence if there is a systematic variation in the measured property. Most sampling schemes employed are a combination of both systematic and random procedures.

A possible outcome of a random selection of points to be sampled is depicted in Fig. 5-2. With this set of sampling points, the rapid change in the property in the lower right would not be detected. Only Point No. 6 is

STATISTICAL PROCEDURES

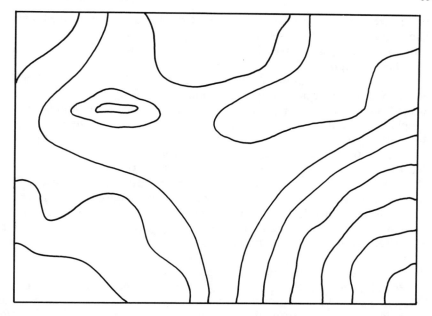

Fig. 5-1. Hypothetical field displaying spatial variability. Lines represent isovalues (equal) of the property.

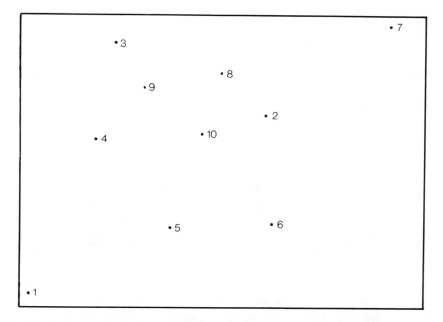

Fig. 5-2. Random sampling scheme, where the location of each point is randomly chosen.

in this area and its observed value would be similar to Points 2, 5 and 7. The average of the 10 observed points might provide an appropriate estimate of the population average, but the scheme would not detect the fact that the assumption of sample independence might not be valid.

A combination sampling scheme is depicted in Fig. 5-3, in which the field is systematically divided into 25 regions. A single point is then randomly located with each of the regions. This scheme is called a random-stratified sampling procedure. Two separate factors have been modified in the scheme, the number of points to be sampled has been increased (25 vs. 10 points) and a restriction has been placed on the location of the points. It is apparent that this scheme will detect most aspects of the variation across the field. The feature that might not be detected is the local extreme in the upper left. We have chosen to increase the number of points sampled in addition to restricting their location in this example, in order to demonstrate that by increasing the number of samples you will not guarantee a full description of the spatial variation. Again this procedure will provide an appropriate estimate of the population average, and will also give some insight into the existence of spatial structure in the soil property.

Finally, consider the three systematic sampling schemes presented in Fig. 5-4, and a slight variation of the procedure in Fig. 5-3. If all samples were collected at the midpoint of the 25 regions in Fig. 5-3, it would represent a grid sampling design. Figure 5-4A, B, and C represent a single transect, orthogonal transects, and radiant transects (or star), respectively. Each of these will detect the spatial structure of the soil property to a different degree.

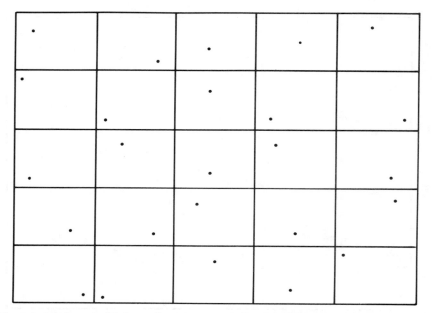

Fig. 5-3. Random stratified sampling scheme. A single point is randomly located within each predefined cell.

STATISTICAL PROCEDURES

In the displayed orientation, the single transect would easily detect the rapid change in the lower left corner. However, had it been oriented differently it might have missed this aspect completely. The orthogonal transects shown would provide a set of observations that would indicate almost no variation in the entire field. The radiant transects and the grid derived from Fig. 5-3 would adequately describe the spatial variation; however, the radiant transect would require significantly fewer samples.

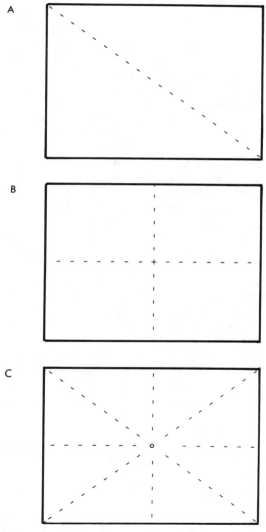

Fig. 5-4. Variations in the transect sampling design: (A) The simplest form with a single transect, (B) two orthogonal transects, and (C) radial transects.

Variation in soil properties range from megascopic (physiographic regions and landforms) to microscopic in scale. For some soil properties it is possible to collect a sample large enough to contain all the spatial variation within one sample volume. Only the expertise of the investigator can provide the needed input to select an appropriate sample size (volume).

PARAMETRIC AND NONPARAMETRIC STATISTICS

Binomial Family of Distributions

Probabilities associated with observations recorded according to the nominal scale (e.g., names) can be estimated by expanding the binomial ($p + q)^n = 1$ where p is the probability of observing a desired taxonomic class within a map unit (a success); and q is the probability of not observing the desired class (a failure). If p is constant for each observation and the observations are independent, the usual estimator of p is $\hat{p} = x/n$ where x is the number of successes and n is the number of observations. The usual estimator of q is $\hat{q} = 1 - \hat{p}$ (Sokal & Rohlf, 1969; Hollander & Wolfe, 1973; Lentner, 1984).

A confidence interval, CI, for p can be obtained from Table A.3 (Hollander & Wolfe, 1973) if sample sizes are <11. For sample sizes >10, the Clopper-Pearson procedure (Hollander & Wolfe, 1973, p. 23) can be used to obtain the CI. The lower limit can be used to make an "at least" statement; and the upper limit can be used to make an "at most" statement about the probability of success. Confidence in p increases as n increases.

If $n\hat{p}$ and $n\hat{q}$ are >5, the number of observations needed to estimate p within a specified error, E, is related by $n = \hat{p}\hat{q} (Z_{\alpha/2}/E)^2$ where $Z_{\alpha/2}$ is given in a table of normal curve areas (Freund & Smith, 1986).

The probability of observing exactly x successes out of n observations is estimated by $P(x) = [n!/x! (n - x)!] p^x q^{n-x}$. Here $P(x)$ estimates the probability of observing a specified number of units possessing a certain soil property or a taxonomic class. The probability of observing x or fewer successes is estimated by the cumulative probability function, $F(v)$, where $F(v) = P(x \le v) = P(x = 0) + P(x = 1) + ... + P(x = v)$. The probability of observing a distribution with $>x$ successes is estimated by $P(x > v) = 1 - F(v)$. Examples of the use of binomial probabilities in the earth sciences are given by Davis (1986), Edmonds and Lentner (1986), Webster (1977), Till (1974), and Griffiths (1967).

A hypergeometric distribution results from sampling a binomial population without replacement. However, there is general agreement that the probabilities associated with the binomial distribution can be used as approximations if less than 5% of a population is sampled (Freund & Smith, 1986).

The chi-square statistic; $\chi^2 = \Sigma(o - e)^2/e$ where o is the observed, and e is the expected frequencies, can be used to compare observed with expected distributions based on theory (Freund & Smith, 1986).

Point Estimators

Mean and Variance

Soil scientists and users of soil survey information are interested in the influence of soil populations on land use. The use of parametric estimators to describe soil populations assumes normality, randomness, and independence.

Parametric mean and variance of populations are estimated by the sample mean, $\bar{x} = 1/n \sum x_i$ (where i = index of sample observations) and sample variance, $s^2 = 1/(n-1)[\Sigma x_i^2 - (\Sigma x_i)^2/n]$. Because s^2 represents squared deviations from x, soil variability is not expressed in terms of the units used to measure the soil properties. The standard deviation, $s = (s^2)^{1/2}$ is a more informative measure because the squaring effect has been reduced, but not eliminated (Lentner, 1984).

Because s depends somewhat on the units of measurement, it is not a consistent estimator of soil variability. This problem can be circumvented by expressing s as a percentage of $\bar{x}$; i.e., the coefficient of variation, CV. Examples of the use of CV in soils research are given by Wilding and Drees (1983), Kahn and Nortcliff (1982), Adams and Wilde (1980), Becket and Webster (1971), and Mader (1963).

The $\bar{x}$ is a variable in random samples and has a distribution, mean, variance, and other parameters. The variance of $\bar{x}$ is estimated by $s_{\bar{x}}^2 = s_x^2/n$. The standard error of $\bar{x}$ is given by $s_{\bar{x}} = s_x/n^{1/2}$. A CI for $\bar{x}$ can be set using the t distribution when σ^2 (population variance) is estimated by s^2 (Lentner, 1984). The CI is $CI_{\bar{x}} = \bar{x} \pm (s/n^{1/2})t_{n-1;1-\alpha/2}$ (α = probability level) where values for $t_{n-1;1-\alpha/2}$ are given in tables for the t-distribution such as Lentner (1984, Table A.6, p. 334), Freund and Smith (1986, Table 3, p. 509) and Ott (1984, Table 2, p. 697). Examples of the estimation of mean, standard deviation, variance, and CI for the mean for selected properties of soils and sediments are given by Edmonds and Lentner (1986), Kahn and Nortcliff (1982), Butler (1980), Webster (1977), Till (1974), McCormack and Wilding (1969), Griffiths (1967), Wilding et al. (1965), and Mader (1963).

The Σx_i, $\bar{x}$, s^2, s, and $s_{\bar{x}}$ are calculated by the UNIVARIATE procedure in SAS (SAS Institute, Inc., 1985a) and by the STATS procedure in SYSTAT (Wilkinson, 1988).

The number of observations included in a sample is important in soil survey research. The number of observations needed to give a $100(1 - \alpha)\%$ CI for $\bar{x}$ can be estimated by $n = (Z_{\alpha/2}/E)^2$ where $Z_{\alpha/2}$ is taken from a table of normal curve areas such as Ott (1984, p. 696); s is the sample standard deviation; and E is the tolerable error. Examples of the estimation of the number samples needed to estimate means within a tolerable error are given by Edmonds and Lentner (1986), Borovskii (1976), Drees and Wilding (1973), Wilding et al. (1965), Aljibury and Evans (1961), and Cline (1944). An example of the number of samples needed to estimate the probability of success, p, for binomial experiments is given by Webster (1977).

Median and Percentile

Nonparametric procedures may be used to estimate the central tendency and variance of soil populations if populations are continuous and observations are independent and random. These procedures can be based on data recorded according to ordinal, interval, or ratio scales. The central concept of nonnormal soil population can be estimated by the sample median, ω, the middle value of the observations arrayed from smallest to largest and denoted as $x_{(1)} \leq \ldots \leq x_{(n)}$. When n is odd or equals $2k + 1$, $\hat{\omega} = x_{(k+1)}$. When n is even or equal to $2k$, $\hat{\omega} = [x_{(k)} + x_{(k+1)}]/2$ (Hollander & Wolfe, 1973).

A $100(1 - \alpha)\%$ CI for ω can be estimated using $C_\alpha = n + 1 - b[(\alpha/2), n, 1/2]$ where $b[(\alpha/2), n, 1/2]$ is given in Table A.2 (Hollander & Wolfe, 1973, p. 262). Observations are ordered from smallest to largest and denoted as $x_{(1)} \leq \ldots \leq x_{(n)}$. The $1 - \alpha$ CI for ω is $\omega_L = x^{(C_\alpha)}$ and $\omega_U = x^{(n+1-C_\alpha)}$ (where L = lower; and U = upper). When samples are larger than 25, C_α is approximately $(n/2) - Z_{\alpha/2}(n/4)^{1/2}$. Because the large sample approximation of C_α is not always an integer, we take the integer closest to C_α for determining ω_L and ω_U. Examples of the use of median and percentiles in soils research are given by Edmonds and Lentner (1987) and Edmonds et al. (1988).

Percentiles can be used to describe variability relative to ω. The 0 (minimum), 1st, 5th, 10th, 25th, 50th (ω), 75th, 90th, 95th, 99th, and 100th (maximum) percentiles are calculated by the UNIVARIATE procedure (SAS Institute, Inc., 1985a). Quartiles, ω, and CI for ω are computed by the GRAPH procedure (Wilkinson, 1988).

Comparison of Two Means or Two Medians

Independent Samples

Independent t-test. Comparison of two independent soil population means is important in soil survey research. Parametric comparison is contingent upon randomness, independence, normality, and equal variances.

Let a common variance for the two populations be represented by σ^2. An unbiased estimator for the variance of the first population is given by s_1^2 with $n - 1$ degrees of freedom and for the second population by s_2^2 with $m - 1$ degrees of freedom. When these two populations are independent, their variances can be combined to give a pooled estimation of σ^2, i.e., $s_p^2 = (SS_1 + SS_2)/(n + m - 2)$ (p = pooled; SS = sum of squares). Information about σ^2 for each sample is contained in s_p^2. Sample estimators of the standard deviations are $s_{x1} = s_p/n^{1/2}$, $s_{x2} = s_p/m^{1/2}$, and $s_d = s_p/(1/n + 1/m)^{1/2}$ (d = difference). The s_p can be used to construct the following t variables, $t_{n+m-2} = (x_1 - \mu_1)/(s_p/n^{1/2})$, $t_{n+m-2} = (x_2 - \mu_2)/(s_p/m^{1/2})$, and $t_{n+m-2} = [(x_1 - x_2) - (\mu_1 - \mu_2)]/[s_p/(1/n + 1/m)^{1/2}]$ (μ = population mean). These t variables have $n + m - 2$ degrees of freedom because s_p, the pooled estimator of s, has these degrees of freedom. These t statistics can be used to test $\mu_1 - \mu_2$ for a difference of δ_0 (δ_0 is a constant difference) and for $\mu_1 = \mu_2$ (Lentner, 1984). Examples of the use of the independent t-test in soils research are given by Edmonds et al. (1986) and Griffiths (1967).

The $100(1 - \alpha)\%$ CI's for μ_1, μ_2, and $\mu_1 - \mu_2$ are $x_1 \pm (s_p/n^{1/2})t_{n+m-2;1-\alpha/2}$, $x_2 \pm (s_p/m^{1/2})t_{n+m-2;1-\alpha/2}$ and $(x_1 - x_2) \pm [s_p/(1/n + 1/m)^{1/2}]t_{n+m-2;1-\alpha2}$, respectively. The t statistic and probability level for testing equality of two independent sample means based on equal and unequal observations are calculated by the TTEST procedure (SAS Institute, Inc., 1985b) and by the STATS procedure (Wilkinson, 1988).

Wilcoxon Rank Sum Test. The Wilcoxon Rank Sum test is a very powerful nonparametric alternative to the t test that tests for differences between ω values for two independent populations without assuming normality (Lentner, 1984). Assumptions basic to this test are random and independent error variables and continuous populations (Hollander & Wolfe, 1973). This test can be based on data recorded according to the ordinal, interval, or ratio scales.

The rank sum statistic, T, is calculated by pooling the n_1 observations from Sample 1 and the n_2 observations from Sample 2 and by ranking them from smallest to largest. The T statistic is the sum of the ranks corresponding to observations from Sample 1. Observations with the same value are assigned the average of the corresponding ranks.

Reject $\omega_1 = \omega_2$ if $T \geq T_{n1, n2, 1-\gamma/2}$ or $T \leq T_{n1, n2, \gamma/2}$. Reject $\omega_1 < \omega_2$ if $T > T_{v1, n2, 1-\gamma}$. Reject $\omega_1 > \omega_2$ if $T < T_{n1, n2, \gamma}$. The γ percentiles are given in Table A.15 (Lentner, 1984, p. 357) for n_1 and $n_2 \leq 20$.

When sample sizes exceed 20, we can make a large-sample rank sum test based on

$$Z = [T - n_1(n_1 + n_2 + 1)/2]/[n_1n_2(n_1 + n_2 + 1)/12]^{1/2}$$

where T is the sum of the ranks of the n_1 observations. Critical Z values are given in tables of the cumulative probabilities of the standard normal distribution (Lentner, 1984, p. 332). An example of the use of the Wilcoxon Rank Sum Test in soils research is given by Edmonds et al. (1988).

The Hodges-Lehmann procedure for estimating the median shift and the Moses procedure for estimating a confidence interval for the median shift are given by Hollander and Wolfe (1973, p. 75 and 78, respectively).

The Mann-Whitney U test, equivalent to the rank sum test, is given by Freund and Smith (1986, p. 460). The Mann-Whitney U statistic and probability level are computed by the NPAR procedure (Wilkinson, 1988). An example of the use of the Mann-Whitney U test in soils research is given by McIntyre and Tanner (1959).

Paired Samples

Paired t-test. Two samples that share common factors are considered to be paired. Examples of paired soil samples are measurements of cation exchange capacity and clay by different methods for the same horizon; and estimation of specific soil minerals, clay, and of base saturation for different horizons of the same pedon. These measurements are separated into two groups; but they are not independent. Therefore, a different test is required. Assumptions basic to this test are normal populations, and random and independent pairs of observations.

The difference between the jth pair is calculated by $d_j = x_{1j} - x_{2j}$. Treat the set of d_j as a single normal population and proceed as in the one-sample case. The mean difference, $d = x_1 - x_2$, is an unbiased estimator of $\mu_d = \mu_1 - \mu_2$ (Lentner, 1984). The variance is estimated by $s_d^2 = 1/(n-1)[\Sigma d_j^2 - (\Sigma d_j)^2/n]$. The paired test statistic is $t_{n-1} = (d - \gamma_0)/(s_d/n^{1/2})$ (γ_0 is a constant). Reject $\mu_d \leq \gamma_0$ if $t_{n-1} \leq t_{n-1;\alpha}$. Reject $\mu_d = \gamma_0$ if $|t_{n-1}| = t_{n-1;1-\alpha/2}$. An example of the use of the paired t-test in the earth sciences is given by Griffiths (1967) and Young (1962).

A 100 $(1-\alpha)$% CI for d is given by $d = (s_d/n^{1/2})t_{n-1;1-\alpha/2}$.

The t statistic and probability level for paired sample means are computed by the MEANS procedure (SAS Institute, Inc., 1985b) and by the STATS procedure (Wilkinson, 1988).

Fisher Sign Test. The sign test is one of the most versatile of the nonparametric procedures for testing for differences between ω values for paired samples (Lentner, 1984). Versatility of the sign test is the result of minimal assumptions. This test assumes that data are recorded according to the ordinal, interval, or ratio scales; observations are dichotomous or a criterion exists for assigning a $+$ or a $-$ sign; and probability of a $+$ or a $-$ sign is equal.

The sign test determines whether the proportion of pairs exhibiting a positive difference is zero. Ties or pairs exhibiting no difference are eliminated from the analysis. Reject $p^+ \leq p^-$ (where p = proportion) if $n^+ \geq 1 - \alpha$ percentage points given in Table A.13 (Lentner, 1984). Reject $p^+ \geq p^-$ if $n^+ \leq \alpha$. Reject $p^+ = p^-$ if $n^+ \leq \alpha/2$ or $n^+ \geq 1 - \alpha/2$.

When sample sizes are greater than 100, the usual Z test can be made using $Z = (n^+ - m/2)/(m^{1/2}/2)$ where n^+ is the number of $+$ signs and m is the number of $+$ and $-$ signs after ties have been removed.

The estimated median difference, θ, associated with the sign test is the median of $(D_i, 1 \leq i \leq n)$ where D_i, is the individual differences ordered from smallest to largest. When n is odd or equals $2k + 1$; $\hat{\theta} = D_{(k+1)}$. When n is even or equals $2k$; $\hat{\theta} = [D_{(k)} + D_{(k+1)}]/2$ (Hollander & Wolfe, 1973).

The D_i and probability level for the Fisher sign test are computed by the NPAR procedure (Wilkinson, 1988).

The Thompson-Savur procedure for estimating a $1 - \alpha$ CI for the median shift is given by Hollander and Wolfe (1973, p. 48).

Wilcoxon Signed-Rank Test. The Wilcoxon signed-rank test is more powerful than the sign test because it uses ranks of the observations in addition to dichotomizing them (Lentner, 1984). Assumptions basic to this test are independent error variables and continuous populations. This test can be based on data recorded according to the ordinal, interval, or ratio scales.

STATISTICAL PROCEDURES

To perform this test, obtain the appropriate difference from the null value or from the pairs, rank the differences ignoring signs, and restore to each rank the corresponding sign. Two types of ties are considered by this test. Observations equal to the null value and pairs with zero differences are discarded. Pairs with equal difference are given average ranks. The test statistic, T, is the sum of the positive ranks. Table A.14 (Lentner, 1984) gives percentage points for this test.

When $n > 25$, the usual one- and two-sided Z tests can be made using

$$Z = [T - m(m + 1)/4]/[m(m + 1)(2m + 1)/24]^{1/2}$$

where T is the sum of the positive signed ranks and m is the number of ranked observations.

The test-statistic and probability level for the Wilcoxon signed rank test are computed by the NPAR procedure (Wilkinson, 1988).

The Hodges-Lehmann procedures for estimating the median shift and the Tukey procedure for estimating a confidence interval for the median shift associated with the Wilcoxon signed-rank test are given by Hollander and Wolfe (1973, p. 33 and 35, respectively).

Comparison of Three or More Means or Medians

Analysis of Variance (One Way)

The parametric means of two or more soil populations can be compared using analysis of variance (ANOVA) if the linear model is appropriate; the experimental units are as homogeneous as possible; the overall mean is a fixed constant, common to all observations; and the residual errors are normally and independently distributed. Deviation from any of these assumptions results in approximate inference (Lentner, 1984).

Because soil scientists have little or no control over natural soil variability, only completely randomized, CR, designs will be discussed. Advantages of the CR designs are flexibility, ease of analysis, and maximum degrees of freedom for estimating residual variability. A disadvantage of the CR designs is that all unrecognized and extraneous variation is included in the residual term resulting in relatively lower power when another design is more appropriate.

The random-effects case is generally applicable to soil survey data because we wish to make inferences about total populations of pedons from which samples were taken. The linear model for the CR design is $x_{ij} = \mu + \tau_i + \epsilon_{ij}$ where x_{ij} is the jth observation from the ith population; μ is the overall mean value; τ_i is the effect of belonging to the ith population; and ϵ_{ij} is the deviation of the (i, j)th observation from the ith population mean (Lentner, 1984).

The ANOVA actually subdivides the total sum of squares, SS_x, into two parts; variation among the populations, P_{xx}, and residual variation, R_{xx}. The T_i population totals are given by Σx_i. The population means, x_i, are given by T_i/n_i. The overall total is given by the sum of the observations for the entire experiment. The overall mean is given by the sum of the observations for the entire experiment divided by the number of observations for the entire experiment.

The sum of squares is estimated by the general correction term, $C = \Sigma(T_i)^2/n$, the population sum of squares, $P_{xx} = \Sigma(T_i)^2/n - C$; the total sum of squares, SS_x, equals the overall total minus C, and the residual or within population sum of squares, $R_{xx} = SS_x - P_{xx}$.

The degrees of freedom associated with the sum of squares are $k - 1$ for the k populations, $n - k$ for the residual or within populations, and $n - 1$ for the total.

The mean squares are the sum of squares divided by their respective degrees of freedom. Mean square for the populations, MS_P, is $P_{xx}/(k - 1)$, and mean square for residual, MS_R, is $R_{xx}/(n - k)$.

Equal populations means can be tested against the alternative that at least two population means are different using $F = MS_P/MS_R$. Reject equal population means if $F \geq F_{k-1; n-1; 1-\alpha}$.

The sum of squares, degrees of freedom, mean squares, F statistic, and probability level are computed for equal numbers of observations by the ANOVA procedure (SAS Institute, Inc., 1985b) and by the MGLH procedure (Wilkinson, 1988). These values for unequal numbers of observations are computed by the GLM procedure (SAS Institute, Inc., 1985b).

The objective of using an ANOVA procedure on experimental results in soil survey is to determine which population means are different. A significant F value indicates a difference between at least two means, but not which of the means. The MEANS procedure (SAS Institute, Inc., 1985b) gives several options for determining a significant difference among means; e.g., Bonferroni t test, Duncan's multiple-range test, Gabriel's multiple-comparison procedure, least significant difference test, and Tukey's studentized range test. The STATS procedure (Wilkinson, 1988) gives Duncan multiple-range test, Tukey's HSD test, and the Newman-Keuls test for determining significant differences among means. Examples of the use of ANOVA in the earth sciences are given by Davis (1986), Edmonds and Lentner (1986), Edmonds et al. (1985), Webster (1977), Wright and Wilson (1979), Till (1974), Ike and Clutter (1968), Griffiths (1967), Wilding et al. (1965), Mader (1963), Aljibury and Evans (1961), and Hammond et al. (1958).

Kruskal-Wallis

If ANOVA assumptions are not valid, the nonparametric Kruskal-Wallis procedure can be used to compare medians for three or more populations. The Kruskal-Wallis procedure is an extension of the two-sample rank sum test (Lentner, 1984). Assumptions basic to this test are independent error variables and continuous populations. The test can be based on data recorded according to the ordinal, interval, or ratio scales.

The data consist of n_i observations from the ith population. The n observations of the k populations being compared are ranked. The observations for each population are replaced by their corresponding ranks. The sum of the ranks for each population is R_i. The Kruskal-Wallis test statistic is

$$H = 12/n(n + 1) \Sigma R_i^2/n_i - 3(n + 1) \qquad i = 1,k.$$

The quantity H has an approximate chi-square distribution with $k - 1$ degrees of freedom and is analogous to MS_P in ANOVA. Reject equal population medians if $H > \chi^2_{(k-1)}$ where $\chi^2_{(k-1)}$ is taken from a table for percentiles of χ^2 distributions; e.g., Table A.7 (Lentner, 1984, p. 335).

A nonparametric multiple comparison procedure based on Kruskal-Wallis rank sums, R_i/N_i, is given by Hollander and Wolfe (1973, p. 124). An example of the use of the Kruskal-Wallis test in soils research is given by Edmonds and Lentner (1987).

The test statistic, probability level, and rank sums for the Kruskal-Wallis test are computed by the NPAR procedure (Wilkinson, 1988).

Analysis of Variance (Nested)

The source and magnitude of soil variability is used to define map units in soil surveys. A nested ANOVA design subdivides soil variability within a map unit into several sources. For example, the variation could be divided among delineations, within map units, among sampling sites (pedons) within delineations, among soil profiles within sampling sites, and among observations; i.e., residual or error variation. The model for such an ANOVA design is $Y_{ijkl} = \mu + D_i + S_{ij} + P_{ijk} + \epsilon_{ijkl}$ where i, j, k and l are index variables; μ is the overall mean; D_i is the effect of belonging to a particular delineation; S_{ij} is the effect of belonging to a particular sampling site; P_{ijk} is the effect of belonging to a particular profile; and ϵ_{ijkl} is the residual or error variance. Assumptions basic to this design are normally and independently distributed components with zero means and respective variances of σ_D^2, σ_S^2, σ_P^2, and σ_ϵ^2 (where ϵ represents the error term).

The variation contributed by delineations within map units, σ_D^2, is estimated by $(MS_D - MS_S)/bcr$ where MD_D and MS_S are mean squares for delineations and sample sites, respectively, taken from an appropriate ANOVA table; and r is the number of replications; c is the number of profiles per sampling site; and b is the number of samples per delineation. The variation contributed by sites within delineations, σ_S^2, is estimated by $(MS_S - MS_P)/cr$ where MS_P is the mean square for profiles. The variation contributed by profiles within sites, σ_P^2, is estimated by $(MS_P - MS_\epsilon)/r$ where MS_ϵ is the residual mean square. The total variance σ_T^2 is estimated by $\sigma_D^2 + \sigma_S^2 + \sigma_P^2 + \sigma_\epsilon^2$. Examples of the use of nested ANOVA in soils research is given by Edmonds and Lentner (1986), Edmonds et al. (1982), and Webster (1977).

The NESTED procedure (SAS Institute, Inc., 1985b) computes the sum of squares, degrees of freedom, mean squares, estimated variance components, and percentage of the total variation contributed by each component in the nested design.

The MGLH procedure (Wilkinson, 1988) computes sum of squares, degrees of freedom, mean squares, F ratios, and probability levels for nested designs.

There is no known parametric counterpart of the nested ANOVA procedure.

GEOSTATISTICS

The term geostatistics is used to describe a set of statistical tools that are extensions of classical statistics with the assumption of sample independence removed. Geostatistical tools are applicable to the measurement of the degree of sample dependence and then to the estimation of population parameters based on the measured sample dependence. The term geostatistics was originally used by a French mining engineer named Matheron. The term and the associated field of statistics represent the combination of geology with statistics. Statistics, as was mentioned at the beginning of this chapter is involved in the study of variability, the randomness that exists in observations. Geologists are involved in the study of the structure of geologic formation. Geostatistics is intended to provide a basis for using both the random and structured features of a set of observations to make scientific estimations.

It is apparent that this group of procedures is well adapted to the study of soils and their variation. The magnitude of scientific literature that has been produced since the late 1970s demonstrates the significance of these tools to soils research. A series of three papers (Burgess & Webster, 1980a,b; Webster & Burgess, 1980) represent some of the early applications of these procedures to the study of soils and are frequently cited for this contribution. These papers present a coherent and fairly thorough description of the application of geostatistics to soil mapping. Descriptions of new applications and innovative developments in the use of geostatistical tools continue to be published, including works such as Wilding and Drees (1983), Trangmar et al. (1985), Webster (1985), and Webster and Oliver (1990), to cite only a few. We refer the reader to these papers for a more comprehensive review of the available literature. Our purpose here will be to give only a cursory overview of these important tools, attempting to stimulate the readers interest in pursuing a more thorough study of the topic.

Regionalized Variables

The tools of geostatistics are based on the theory of regionalized variables that was derived from work by Krige (1951) and developed into the present theory by Matheron (1971). It was the observed continuity of geologic properties that led to the development of this theory. In its basic form regionalized variable theory says that the variable changes from point to point in space in a continuous manner that may be described by some mathematical function. The regionalized variable is a realization of that function at some point. There are numerous sources that give a complete description of the theory of regionalized variables, and the reader is referred to those for further information (David, 1977; Journel & Huijbregts, 1978; Clark, 1979; Royle, 1980).

Semivariogram

The basic tool of geostatistics is the semivariogram. Some confusion has developed around the name of this function. It is not the place of this chapter to contribute either to the solution of the confusion or to its proliferation. We will choose to use the contracted term variogram in place of semivariogram, only for convenience.

The variogram is defined by

$$\gamma(\mathbf{h}) = 1/2 \; E\{[Z(x_i) - Z(x_i + \mathbf{h})]^2\}$$

where E is the expected value of the function within the braces ({}); $Z(x)$ is the regionalized variable at x; and $\mathbf{h}$ is the vector of some length and direction. In practice the regionalized variable is not known; but observations provide us with an estimate of its value. The estimated value is a random variable represented by $z(x)$, and the estimated or "raw" variogram is defined by

$$\gamma^*(\mathbf{h}) = [2N(\mathbf{h})]^{-1} \Sigma \; [z(x_i) - z(x_i + \mathbf{h})]^2 \qquad i = 1, N(\mathbf{h})$$

where $N(\mathbf{h})$ is the number of pairs of points separated by $\mathbf{h}$. A fact that is not obvious in this definition is that $\mathbf{h}$, as a vector, has both magnitude and direction. This suggests the implicit assumption that the regionalized variable is isotropic. It is generally the case that soil properties are anisotropic, meaning that the variability depends on both distance and direction. Therefore it is important to test for anisotropy in the analysis of soil properties.

In the theory of regionalized variables, it is assumed that the property is stationary. Simply stated this implies that the mean and the variance of the property are invariant on translation and that the variance is finite. This is stated functionally

$$E[z(x) - z(x + \mathbf{h})] = 0$$

where the points $z(x)$ and $z(x + \mathbf{h})$ are any two points in the region sampled that are separated by $\mathbf{h}$. Constant variance applies

$$\text{Var} \; [z(x) - z(x + \mathbf{h})] = E\{[z(x) - z(x + \mathbf{h})]^2\} = 2\gamma(\mathbf{h})$$

Taken together these form the intrinsic hypothesis of stationarity required for the variogram. The intrinsic hypothesis is only violated when there are

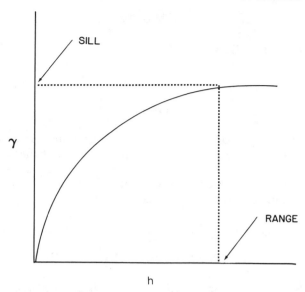

Fig. 5-5. General form of the variogram indicating the range of influence and the sill.

obvious drifts or trends in the regionalized variable. Procedures for removing trends are discussed in detail by Webster and Burgess (1980).

Further interpretation of the spatial structure of the variable of interest is accomplished by modeling the variogram. A few of the commonly used models are linear model, power functions, spherical model, exponential model, Gaussian model, and cubic model. These models may be used directly or may be combined into more complex models. The fitting of the raw variogram to one of these models is one of the least vigorous procedures in the area of geostatistics. This is often accomplished by trial and error. However, further analysis is very robust with respect to the choice of the model for the variogram.

There are a few general characteristics of the variogram that should be mentioned. The classical shape of the variogram is shown in Fig. 5-5. The variogram is expected to penetrate the origin, increase to some value, and stabilize. The value at which it stabilizes is called the sill and is equal to the sample variance. The distance at which the variogram reaches the sill is called the range. The range provides a direct determination of sample independence. Samples separated by a distance greater than the range are independent, while those within the range are spatially dependent. A more complete description of the variogram is given in Upchurch et al. (1988).

Kriging

The ultimate goals of the study of spatial variability is to estimate the value of the regionalized variable at points that have not been sampled. The process of interpolation between sampled points using the spatial structure described by the variogram is called kriging. The name is derived from the name of the mining engineer, Krige, who first developed the procedure.

The mathematical development of the kriging system is well documented in the literature and only a few points need to be covered for an adequate understanding of the system. The estimated value of the variable (y_0^*) is calculated by

$$y_0^* = \Sigma \lambda_i z(x_i) \qquad i = 1, n$$

where the values for lambda (λ) represent the weights associated with the measured values, z, at the point x, in order to estimate the value at the point of interest, x_0. The weights are determined such that the estimate is the best, linear, unbiased estimate.

In order for the estimator to be unbiased, the weights must sum to 1, i.e.,

$$\Sigma \lambda_i = 1 \qquad i = 1, n.$$

Without going into the complete mathematical details, the kriging system is defined by

$$\Sigma \lambda_j \gamma(x_i, x_j) + \mu = \gamma(x_i, V) \qquad j = 1, n \qquad \text{for all } i.$$

The kriging variance or estimation variance is defined by

$$\sigma_k^2 = \Sigma \lambda_i \gamma(x_i, V) - \gamma(V, V) + \mu \qquad i = 1, n.$$

These systems are best solved numerically in their matrix form

$$\mathbf{FX} = \mathbf{B}$$

where $\mathbf{F}$ contains information about the sampled points; $\mathbf{X}$ contains the weights; and $\mathbf{B}$ contains information about the location to be estimated. The exact form of the matrices can be obtained from Vieira et al. (1983). The system described is the general system of equations for kriging, and has been called block kriging. This name is derived from the fact that the estimated value represents a region or block, of volume V, rather than a point.

The kriging interpolation system has an advantage over traditional interpolation methods from two standpoints. The kriging estimation is based on the identified spatial structure rather than some predefined procedure of weighting. Of most importance the kriging system provides both the estimate and a value for the error associated with that estimate. Using the estimation variance, a confidence interval can be calculated for the interpolated value. The previous section dealing with traditional statistical procedures provides the details of appropriate tests for the differences between population means. These tests are directly applicable to testing the differences between interpolated values provided by the kriging system.

An extension of the kriging system that is being used in the study of soils is called cokriging. This would apply when two separate but related variables have been measured at several locations and one of the variables has been measured at additional locations. The information from the oversampled variable is used in the estimation of the undersampled variable. If the variables are related such that their spatial structures are correlated, the information from both variables can be used in the estimation process and will yield smaller estimation variances. The details of cokriging are described in Vieira et al. (1983).

Time Series Analysis

In many experiments the data collected are most appropriately represented as sequences of data. That is, they are characterized by their position along a single line. The position, in the sequence or the line, at which the data point occurred is important. Traditionally this type of data has been analyzed using procedures associated with time-series analysis. These procedures are appropriate for the analysis of data that have a single positional characteristic. Several of these procedures are applicable to the analysis of data collected over time or space.

In the analysis of sequences of data the first and overriding question that should be asked is, Are the observations random? If the observations are not random they must contain evidence of a pattern or a trend. The second question is then apparent, What is the form of the trend or pattern? For example, Are there repetitions or cycles in the observations? If there are repetitions or cycles, What is the frequency of the cycle? These all lead to the two practical questions, Can predictions, such as interpolation between data points or extrapolation beyond the observations be made? Are two separate sequences similar?

The first tool to be described is autocorrelation and the associated graphical display, the correlogram. This tool is useful in detecting repetitions and trends in a sequence of observations. The sequence is compared to itself at successive positions to determine the degree of similarity at intervals. A requirement for this procedure is that the observations must be in a sequence with a constant separations distance. This separation distance is referred to as the lag, and comparisons are then made at integer multiples of the lag. Since the lag is a specific, known distance, the distance at which the comparisons are made can be determined; although this value is not actually used in the calculations.

An estimate of the autocorrelation at lag L can be calculated by (Davis, 1978)

$$r_L = \frac{[(n - L)(\Sigma\, z_i z_{i+L})\, \Sigma\, z_i \Sigma z_{i+L}]/(n - L)(n - L - 1)}{[n\Sigma z_i^2 - (\Sigma z_i)^2]/n(n - 1)}$$

where n is the number of points in the sequence and the z's are the observed data. The correlogram is constructed by plotting the autocorrelation for each lag vs. its lag. The r_L is constrained between -1 and 1, which represents a perfect negative and positive correlation, respectively. A value of 0 indicates no correlation at a given lag.

At a lag of 0 the autocorrelation is 1; that is every point is perfectly correlated with itself. In general r_L declines as the lag increases with its behavior determined by the structure of the data. Davis (1986) gives an excellent discussion of the behavior of the autocorrelation of idealized time series.

Once it has been established through the autocorrelation that a trend or pattern exists in a sequence of observations, the spectral density function can be used to further describe the pattern. The spectral density function is a Fourier transform of the autocorrelation function. The spectral density function can reveal periodicities in the data and the associated frequency. A more thorough description of the spectral density function is given in Upchurch et al. (1988) and the complete details can be found in other publications such as Box and Jenkins (1976).

The last tool that we will mention in this group is the cross-correlation function. Comparisons between separate sequences of data can be made using this function. Conceptually the two sequences are moved past each other and the degree of correlation is determined as the match point is changed. As implied, this is a procedure for comparing two sequences of data, therefore a test for significance can be made. The significance of the cross-correlation is assessed by a t-test. The t value is defined as

$$t = r_m[(n^* - 2)/(1 - r_m^2)]^{1/2}$$

where r_m is the cross-correlation at match position m and n^* is the number of overlapping positions in the sequence. This value can be compared to a t value with $(n^* - 1)$ degrees of freedom, to determine if the cross-correlation of the two sequences is significant. The calculation of the cross-correlation function and interpretation of the test is described in some detail by Davis (1986).

SUMMARY

This has been a rather brief and cursory discussion of the statistical procedures available and used in investigation of soils. However, it should provide a direction for further study of the literature by those interested in the subject. We have attempted to introduce the procedures and their application without giving many of the operational details, hoping to provide a simple guide to appropriate procedures with specific objectives.

Two general groups of procedures have been discussed, those in which the assumption of sample independence is appropriate and those in which this assumption is investigated and sample dependence is exploited. The distinction between these groups is somewhat artificial. The group of procedures we have placed in the category of geostatistics is in fact a logical extension of the more traditional parametric and nonparametric statistics familiar to all researchers.

REFERENCES

Adams, J.A., and R.H. Wilde. 1980. Comparison of variability in a soil taxonomic unit with that of the associated soil mapping unit. Aust. J. Soil Res. 18:285–297.

Aljibury, F.K., and D.D. Evans. 1961. Soil sampling for moisture retention and bulk density measurements. Soil Sci. Soc. Am. Proc. 25:180–183.

Becket, P.H.T., and R. Webster. 1971. Soil variability: A review. Soils Fert. 34:1–15.

Borovskii, V.W. 1976. Experiences in the study of the variability of certain properties of the soils of South Kazakhstan. p. 42–54. In W.M. Fridland (ed.) Soil combinations and their genesis. Amerind Publ. Co. Pvt. Ltd., New Delhi.

Box, G.F.P., and G.M. Jenkins. 1976. Times series analysis. Holden-Day, San Francisco, CA.

Burgess, T.M., and R. Webster. 1980a. Optimal interpolation and isarithmic mapping soil properties. I. The semivariogram and punctual kriging. J. Soil Sci. 31:315–333.

Burgess, T.M., and R. Webster. 1980b. Optimal interpolation and isarithmic mapping soil properties. II. Block kriging. J. Soil Sci. 31:333–341.

Butler, B.E. 1980. Soil classification for soil survey. Clarendon Press, Oxford, United Kingdom.

Clark, I. 1979. Practical geostatistics. Applied Sci. Publ., London.

Cline, M.G. 1944. Principles of soil sampling. Soil Sci. 58:275–288.

David, M. 1977. Geostatistical area reserve estimation. Elsevier Sci. Publ. Co., Amsterdam.

Davis, J.C. 1978. Statistics and data analysis. John Wiley & Sons, Toronto.

Davis, J.C. 1986. Statistics and data analysis, 2nd ed. John Wiley & Sons, New York.

Drees, L.R., and L.P. Wilding. 1973. Elemental variability within a sampling unit. Soil Sci. Soc. Am. Proc. 37:82–87.

Edmonds, W.J., J.C. Baker, and T.W. Simpson. 1985. Variance and scale influence on classifying and interpreting soil map units. Soil Sci. Soc. Am. J. 49:957–961.

Edmonds, W.J., P.R. Cobb, and C.D. Peacock. 1986. Characterization and classification of seaside-salt-marsh soils on Virginia's eastern shore. Soil Sci. Soc. Am. J. 50:672–678.

Edmonds, W.J., S.S. Iyengar, L.W. Zelazny, M. Lentner, and C.D. Peacock. 1982. Variability in family differentia of soils in a second-order soil survey mapping unit. Soil Sci. Soc. Am. J. 46:88–93.

Edmonds, W.J., and M. Lentner. 1986. Statistical evaluation of the taxonomic composition of three map units in Virginia. Soil Sci. Soc. Am. J. 50:997–1001.

Edmonds, W.J., and M. Lentner. 1987. Soil series differentiae selected by discrimination analysis based on ranks. Soil Sci. Soc. Am. J. 51:716–721.

Edmonds, W.J., D.D. Rector, N.O. Wilson, and T.L. Arnold. 1988. Evaluation of relationships between oak site indices and properties of selected dystrochrepts. Soil Sci. Soc. Am. J. 52:204–209.

Freund, J.E., and R.M. Smith. 1986. Statistics a first course. 4th ed. Prentice-Hall, Inc., Englewood Cliffs, NJ.

Griffiths, J.C. 1967. Scientific method in analysis of sediments. McGraw-Hill Book Co., New York.

Hammond, L.C., W.L. Pritchett, and C. Chew. 1958. Soil sampling in relation to soil heterogeneity. Soil Sci. Soc. Am. Proc. 22:548–552.

Hollander, M., and D.A. Wolfe. 1973. Nonparametric statistical methods. John Wiley & Sons, New York.

Ike, A.F., and J.L. Clutter. 1968. The variability of forest soils of the Georgia Blue Ridge Mountains. Soil Sci. Soc. Am. Proc. 32:284–288.

Journel, A.G., and Ch. J. Huijbregts. 1978. Mining geostatistics. Academic Press, London.

Kahn, M.A., and S. Nortcliff. 1982. Variability of selected soil micronutrients in a single soil series in Berkshire, England. J. Soil Sci. 33:763–770.

Krige, D.G. 1951. A statistical approach to some mine valuations and allied problems at the Witwatersrand. M.S. University of Witwatersrand, South Africa.

Lentner, M. 1984. An introduction to applied statistics. 2nd ed. Valley Book Co., Blacksburg, VA.

Mader, D.L. 1963. Soil variability — A serious problem in the soil-site studies in the northeast. Soil Sci. Am. Soc. Proc. 27:707–709.

Matheron, G. 1971. The theory of regionalized variables and its applications. Les Cahiers du Centre de Morphologie Mathematique 5, Centre de Geosstaatisstique, Fontainebleau, France.

McCormack, D.E., and L.P. Wilding. 1969. Variation of soil properties within mapping units of soils with contrasting substrata in Northwestern Ohio. Soil Sci. Soc. Am. Proc. 33:587–593.

McIntyre, D.S., and C.B. Tanner. 1959. Anormally distributed soil physical measurements and nonparametric statistics. Soil Sci. 88:133–137.

Ott, L. 1984. An introduction to statistical methods and data analysis. 2nd ed. Duxbury Press, Boston.

Royle, A.G. 1980. Geostatistics. McGraw-Hill, New York.

SAS Institute, Inc. 1985a. SAS user's guide: Basics, version 5 ed. SAS Inst., Inc., Cary, NC.

SAS Institute, Inc. 1985b. SAS user's guide: Statistics, version 5 ed. SAS Inst., Inc., Cary, NC.

Sokal, R.R., and F.J. Rohlf. 1969. Biometry. W.H. Freeman and Co., San Francisco.

Till, R. 1974. Statistical methods for the earth scientist: An introduction. John Wiley & Sons, New York.

Trangmar, B.B., R.S. Yost, and G. Uehara. 1985. Application of geostatistics to spatial studies of soil properties. Adv. Agron. 38:45–94.

Upchurch, D.R., L.P. Wilding, and J.L. Hatfield. 1988. Methods to evaluate spatial variability. p. 201–229. *In* L.R. Hossner (ed.) Reclamation of surface-mined lands. Vol. II. CRC Press, Inc., Boca Raton, FL.

Vieira, S.R., J.L. Hatfield, D.R. Nielsen, and J.W. Biggar. 1983. Geostatistical theory and application to variability of some agronomical properties. Hilgardia 51:3.

Webster, R. 1974. Quantitative and numerical methods in soil classification and survey. Clarendon Press, Oxford, United Kingdom.

Webster, R. 1985. Quantitative spatial analysis of soil in the field. p. 1–70. *In* B.A. Steward (ed.) Advances in soil science 3. Springer-Verlag, New York.

Webster, R., and T.M. Burgess. 1980. Optimal interpolation and isarithmic mapping of soil properties. III. Changing drift and universal kriging. J. Soil Sci. 31:505–525.

Webster, R., and M. Oliver. 1990. Statistical methods in soil and land resource survey. Oxford Univ. Press, Oxford, United Kingdom.

Wilding, L.P., and L.R. Drees. 1983. Spatial variability and pedology. p. 83–116. *In* L.P. Wilding et al. (ed.) Pedogenesis and soil taxonomy. I. Concepts and interactions. Elsevier, New York.

Wilding, L.P., R.P. Jones, and G.M. Schafer. 1965. Variation of soil morphological properties within Miami, Celina, and Crosby mapping units in west-central Ohio. Soil Sci. Soc. Am. Proc. 29:711–717.

Wilkinson, L. 1988. SYSTAT: The system for statistics. Systat, Inc., Evanston, IL.

Wright, R.I., and S.R. Wilson. 1979. On the analysis of soil variability, with an example from Spain. Geoderma 22:297–313.

Young, K.K. 1962. A method of making moisture desorption measurements on undisturbed soil samples. Soil Sci. Soc. Am. Proc. 26:301.

6 A Comparison of Statistical Methods for Evaluating Map Unit Composition

S. C. Brubaker

USDA-ARS
Lincoln, Nebraska

C. T. Hallmark

Texas A & M University
College Station, Texas

ABSTRACT

Map unit composition has traditionally been determined by transecting randomly selected delineations of the map unit in question, recording the soil series at each point on each transect, and calculating the average proportion of each soil that occurs within the map unit. Confidence intervals about the mean for each soil are then determined using either the Student's t-distribution or a binomial method. Data from randomly selected transects were analyzed by both methods and the results were compared. The influence of nonuniform sampling densities, and variations in the number of points sampled in each delineation were also examined. Results indicate that for small sample sizes the binomial method gave narrower confidence intervals; but for large samples the two methods were equivalent. Due to the large number of samples required for geostatistics, its use in evaluating map unit composition is limited.

The additional pressure on our land resource brought about by increasing population, urbanization, and soil erosion has significantly increased the demand for soil surveys and the information they provide. In addition, the use of soil survey information for a variety of nonagricultural land uses and its use as a regulatory tool have resulted in an increasing demand for greater detail and more specific information. Users want and need confidence limits, probabilities, and frequency analyses on the composition of map units and information on how the inclusions within a given map unit influence the interpretations (Miller, 1978). In order to provide this information to the users, some authors have suggested that a table be included in each soil survey

---

Copyright © 1991 Soil Science Society of America, 677 S. Segoe Rd., Madison, WI 53711, USA. *Spatial Variabilities of Soils and Landforms.* SSSA Special Publication no. 28.

that would provide information about the composition of each map unit (Amos & Whiteside, 1975; Nordt et al., 1991). Therefore, an efficient method for evaluating map unit composition that would provide the needed information and could easily be incorporated into the soil survey program should be established (Hajek, 1982).

The objectives of this study were to (i) review the current literature on methods for evaluating map unit composition, (ii) compare the various methods with generated data and data from a previous study on map unit composition, and (iii) assess the applicability of geostatistical methods for evaluating map unit composition.

LITERATURE REVIEW

A soil map unit, as defined by the Soil Survey Staff (1983), is a collection of soil areas delineated in mapping with each soil area being made up of one or more contiguous pedons. Each map unit is named for and contains a limited number of principal soils and has a limited range of proportions and patterns of these soils. However, even though soil map units are named in terms of the dominant taxonomic phases present, nearly all contain areas of soil that are not identified in the map unit name. These mapping inclusions are grouped into two categories, those which are similar to the named soil, or soils, and those which are dissimilar. The primary goal in evaluating map unit composition is to identify the proportion of the map unit that is comprised of the named soil, or soils, and to describe the abundance and nature of the mapping inclusions.

Map unit composition has traditionally been quantified by transecting selected delineations of the map unit being studied and determining at each point on the transect whether or not the soil is the same as, or is similar to, the named series (Powell & Springer, 1965; Amos & Whiteside, 1975; Steers & Hajek, 1979; Bigler & Liudahl, 1984; Hopkins et al., 1987; Nordt et al., 1991). The percentage of points on each transect that were identified as such was then recorded and the average over all transects used as an estimate of the compositional purity of the map unit. Confidence intervals were then calculated using either the Student's t-distribution (Hajek & Steers, 1981; Bigler & Liudahl, 1984) or a binomial method (Edmonds & Lentner, 1986; Nordt et al., 1991).

Although the transect method has been used most frequently for studying map unit composition, other sampling designs such as stratified random sampling (Wilding et al., 1965; Beckett & Burrough, 1971; Bascomb & Jarvis, 1976) have also been used. These various methods have been described in detail by several authors including Wilding & Drees (1983), Upchurch et al. (1988), and Burrough (1991).

In addition to differences in sampling and analytical techniques, the various studies of map unit composition have also differed in the amount of dissimilarity among soils that was allowed in the definition of the map unit. A majority of the earlier studies focused primarily on the taxonomic purity of the map unit. Their primary objective was to determine what proportion of the soils within the map unit were in the same taxonomic class as the named soil, or soils. Although the *Soil Survey Manual* (Soil Survey Staff, 1951) required map unit purities of 85%, the taxonomic purity found in many of these studies was reported to be 50% or less (Wilding et al., 1965; McCormack & Wilding, 1969; Amos & Whiteside, 1975; Edmonds & Lentner, 1986; Hopkins et al., 1987). Recent studies not only examined the taxonomic purity of the map units at the various levels of the classification system, but also examined the interpretive purity of the map units (West et al., 1981; Nordt et al., 1991). By including soils that were taxonomically dissimilar but had similar interpretations in the map unit, Nordt et al. (1991) were able to increase the map unit purity for two of the four map units they studied to greater than 80%. Overall, the addition of similar soils to the map unit increased the map unit purity by 30 to 35%. This result indicates that it may be easier to obtain a high degree of interpretive purity within a given map unit than it is to achieve a high degree of taxonomic purity. Since most users would be more interested in the interpretive purity of the map units, this information should probably be included in the soil survey along with or in lieu of information on the taxonomic purity of the map units.

Sampling Considerations

Since it would be impossible to sample or classify every pedon within a given map unit delineation, in order to estimate the composition of the map unit it is necessary first to estimate the composition of a selected number of delineations and then use those estimates to estimate the composition of the entire map unit. This is a classical situation of sampling in stages. The primary sampling units are the delineations of the map unit; while the sampling units at the second stage are either cores from individual pedons, if random sampling is used to estimate the composition of the delineations, or individual transects, if transects are used to estimate the composition of the delineation. In the latter case, the cores from individual pedons along the transect would be the sampling unit at the third stage.

When used in conjunction with analysis of variance, this sampling technique allows for estimation of the amount of variation at each level; i.e., it provides estimates of the amount of variation within delineations as well as the amount of variation between delineations. In addition, this partitioning of the variance provides a way to estimate the number of samples needed at each level to obtain a more precise estimate of the population mean and provides a method for determining the efficiency of the sampling scheme for the allocation of resources. These methods were used by Wilding et al. (1965) and McCormack and Wilding (1969) to describe the variation of soil properties within map units in Ohio. A detailed discussion of this sampling technique and the associated analysis, as well as a discussion of its application for determining the allocation of resources is provided by Snedecor and Cochran (1967).

To insure that the estimate is unbiased, the delineations to be studied should be selected by either a random sampling technique or a stratified random sampling technique. Often it is more desirable to stratify the delineations either by size or by areal distribution in order to insure that the delineations selected accurately represent the map unit. Several authors have discussed stratification techniques including Snedecor and Cochran (1967), Wang (1982), and de Gruijeter and Marsman (1985).

After the delineations have been selected, several different methods for estimating their composition can be used. These include both random and systematic techniques. The most important analytical consideration, regardless of which sampling technique is chosen, is that the sampling density is the same for all delineations. In addition, if the samples will also be used to estimate the mean and variability of selected soil properties, the samples should be spaced sufficiently far apart to insure that they are nearly independent relative to the properties being studied.

Two questions that often arise in studies such as these are, How many samples should be taken? and How far apart should the samples be? Of the two questions, the first is by far the most easily answered. Sample number is most commonly determined using the following equation

$$N = \frac{t_{\alpha,\nu}^2 \times s^2}{\Delta^2}$$ [1]

where N is the number of samples; $t_{\alpha,\nu}$ is taken from the Student's t-distribution, s^2 is the estimated population variance; and Δ is the desired margin of error. When multiple-stage sampling is used, the methods discussed above can be applied to determine the optimum number of samples needed at each stage of sampling to estimate the mean with a specified level of accuracy.

As far as sample spacing is concerned, Wang (1982) discussed the selection of observation intervals for studies of map unit composition and the various factors that influence the choice. He also provided two arbitrary methods for determining sample spacing along transects. Although both methods have been used in the past, the second method (where observation interval is chosen to be one-half the length of the shortest available transect), is the better method in that it will provide a uniform sample spacing for all delineations and; therefore, the sampling density will be the same for all delineations. The other method, where sampling interval is set to one-tenth the length of the transect, provides the same number of observations per transect but will result in a variable sampling density that will essentially give more weight in the analysis to the shorter transects and less weight to the longer transects.

Recently, new statistical methods developed for the analysis of spatially dependent variables and known collectively as "geostatistics" have been applied to the study of soil variability with some success. One application of geostatistics to the study of map unit composition is to determine the sample spacing necessary to insure that samples are nearly independent relative to the soil property being studied. Brubaker (1989) generated semivariograms for several soil properties for selected upland soils in Brazos

County, TX. These properties included surface and solum thickness; soil pH, extractable bases, and percentage sand, silt, and clay in the upper 15 cm of the A-horizon, the upper 50 cm of the argillic horizon and at the critical depth for determining base saturation for classification. The estimated range of influence for those properties that exhibited spatial dependence ranged from 20 to 150 m; however, a majority of the values were between 30 and 90 m and the overall average was approximately 65 m. Based on these results, he suggested that, for the soils studied, samples should be taken at least 65 m apart in order to insure that they are nearly independent relative to most of the soil properties examined. The primary disadvantages of using geostatistics to determine sample spacing are the large number of samples required to generate the semivariograms and the fact that the results are essentially site specific. However, for high density soil surveys that require greater precision in the estimates of various soil properties the procedure would probably be worthwhile.

Statistical Methods

Two different methods of analysis are presently used to estimate the map unit composition based on the data collected. One is based on the binomial distribution, which is a discrete distribution; and the other is based on the normal distribution, which is a continuous distribution. "The binomial distribution gives the probability that 0, 1, 2, ..., n members of a sample of size n will possess some attribute when the sample is a random sample from a population in which a proportion, p, of the members possess this attribute" (Snedecor & Cochran, 1967). In evaluating map unit composition, we are interested in estimating the proportion, p, of the soils in the map unit that are the same as, or are similar to, the named soil, or soils. The sample estimate of the proportion is given by $\hat{p} = r/n$ where r is the number of soils in the sample that are the same as, or are similar to, the named soil, or soils, and n is the sample size. A confidence interval about p is given by the following equation

$$\hat{p} \pm Z_\alpha \left[\frac{\hat{p}\hat{q}}{n}\right]^{1/2} \qquad [2]$$

where $\hat{p}$ is the estimated proportion of soils within the map unit that are the same as, or are similar to, the named soil, or soils; $\hat{q} = (1 - \hat{p})$; n is the total number of pedons sampled; and Z_α is the value of the standard normal variate for which the probability of a greater value is equal to α. This confidence interval is based on the fact that the shape of the binomial distribution is nearly normal when n is large. For smaller values of n, the same confidence interval with a correction for continuity is given by the equation

$$\hat{p} \pm \left[Z_\alpha \left(\frac{\hat{p}\hat{q}}{n}\right)^{1/2} + \frac{1}{2n}\right] \qquad [3]$$

A discussion on the use of the binomial method for estimating map unit composition and a graphical method for determining the confidence intervals and estimating the number of samples needed to obtain a given level of accuracy are provided in several articles by Arnold (1979, 1981).

The binomial method considers each pedon to be a separate trial with the result being either it is or is not the same as, or is or is not similar to, the named soil, or soils. The results in this case fit a discrete distribution; i.e., the number of pedons that are the same as, or are similar to, the named soil, or soils, based on n trials has to be an integer between 0 and n. An alternate way of looking at the data would be to consider each set of n pedons a separate trial. In this case, the result assigned to a given trial would be the percentage of the pedons in the sample that are the same as, or are similar to, the named soil, or soils. The results from several trials would form a continuous distribution about the mean. Assuming that the samples are from a normal distribution, confidence intervals about the mean can be calculated with the standard normal deviate as was done in the binomial method. This is satisfactory if an adequate number of trials is run; however, for small numbers of trials (<30), it can often produce incorrect results. Therefore, when fewer than 30 trials are run, a t-statistic is generally used in place of the standard normal deviate. Confidence intervals using the t-statistic are calculated using the following equation

$$\bar{x} \pm \frac{t_{\alpha,a-1} \times s}{(a)^{1/2}} \qquad [4]$$

where $\bar{x}$ is the estimated mean; $t_{\alpha,a-1}$ is the value of the t-statistic at the $(1 - \alpha)$ confidence level with $a - 1$ degrees of freedom; s is the estimated standard deviation; and a is the number of transects or delineations sampled.

The major advantage of the latter method for calculating map unit composition is that it allows us to estimate the amount of variability within delineations, providing that more than one set of samples is taken from each delineation, as well as the amount of variability between delineations. The binomial method, on the other hand, provides no estimate of the amount of variation between delineations within the map unit.

The primary disadvantage associated with using the second method to estimate map unit composition is that it can result in biased estimates if care is not taken to account for differences in the size of delineations and the associated difference in the number of samples taken within each delineation. When the binomial method is used, each individual pedon carries the same amount of weight and therefore the size of a delineation, or the number of samples taken from it, would not influence the estimate; however, care should still be taken to insure that the sampling density is the same for all delineations.

Although the two methods just described are the most commonly used, the recent use of geostatistical methods to describe the spatial distribution of selected soil properties (Burgess & Webster, 1980; Yost et al., 1982; Laslett et al., 1987; Oliver & Webster, 1987; and Gaston et al., 1990) suggests that perhaps these methods could also be used to describe the distribution of soils within a delineation. One advantage of this method is that it not only provides an estimate of the value of a specific soil property at a given point or over a given area; but it also provides an estimate of the error associated with that estimate. There are two major disadvantages; however, that would preclude the routine application of geostatistical methods for evaluating map unit composition. The primary disadvantage is that this method is sample intensive, i.e., it requires a large number of samples within the area being described to accurately estimate the semivariogram. The second disadvantage is that these methods are site specific, and results have limited use outside of the area sampled.

METHODS

The two methods of analysis were compared by using a technique similar to that used by White (1966) to test the validity of the transect method for evaluating map unit composition. In this case, 12 blocks consisting of 400 cells each were drawn on graph paper. Within each block, half of the cells were shaded with the distribution of the shaded cells being different for each block (Fig. 6-1). In terms of map unit composition, each block represents a different delineation from a map unit composed of equal amounts of two different soils. One advantage of using this technique for comparing the two methods is that the actual proportion of each block that is shaded is known.

To compare the two analytical methods and examine the influence of different sampling techniques on the results from each method, two experiments were conducted. In the first experiment, 10 points were selected using a random number generator on a microcomputer from each of 2 through 10 randomly selected blocks. The experiment was replicated three times, and the data are presented in Table 6-1. Comparison of the half-width of the

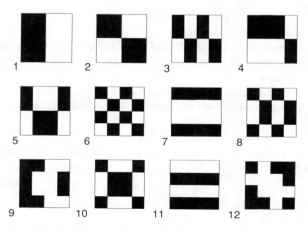

Fig. 6-1. Blocks used to generate the data for method comparison.

Table 6-1. The estimated proportion of shaded area in the blocks and the associated standard deviation based on n randomly selected samples from each of a randomly selected blocks.

a	an†	Proportion	s_1‡	s_2§
		Replication 1		
2	20	0.550	0.354	0.111
3	30	0.367	0.306	0.088
4	40	0.425	0.150	0.078
5	50	0.540	0.089	0.070
6	60	0.483	0.160	0.065
7	70	0.557	0.162	0.059
8	80	0.563	0.106	0.055
9	90	0.522	0.148	0.053
10	100	0.470	0.189	0.050
		Replication 2		
2	20	0.450	0.071	0.111
3	30	0.500	0.100	0.091
4	40	0.450	0.173	0.079
5	50	0.540	0.114	0.070
6	60	0.567	0.163	0.064
7	70	0.386	0.090	0.058
8	80	0.588	0.125	0.055
9	90	0.511	0.145	0.053
10	100	0.560	0.165	0.050
		Replication 3		
2	20	0.400	0.000	0.110
3	30	0.533	0.058	0.091
4	40	0.575	0.250	0.078
5	50	0.500	0.158	0.071
6	60	0.500	0.155	0.065
7	70	0.486	0.135	0.060
8	80	0.575	0.149	0.055
9	90	0.544	0.174	0.053
10	100	0.440	0.171	0.050

† $n = 10$ for each block.
‡ s_1 = Standard deviation from the classical method.
§ s_2 = Standard deviation from the binomial method.

confidence intervals and the estimated standard deviations for each method were then made by using a paired t-test (Table 6-2). In addition, a visual comparison of how the average confidence intervals, the average half-widths of the confidence intervals, and the average estimated standard deviations from each method were influenced by sample size was made by plotting them (Fig. 6-2 to 6-4).

Table 6-2. Statistical comparison of the confidence limits and standard deviations from each method using the paired t-test.

Variable	Mean difference	t	Probability of a greater t
Confidence limits	0.168	1.510	0.1430
Standard deviations	0.080	5.670	0.0001

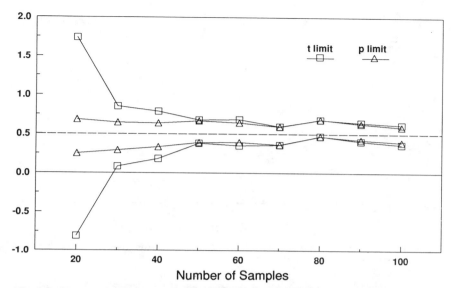

Fig. 6-2. A comparison of the confidence intervals calculated using the *t*-statistic with those calculated by the binomial method.

To examine the influence of the number of samples taken from each block on the results from each analytical method, 5, 10, and 15 points were selected at random from each of 2 through 10 randomly selected blocks (Table 6-3). The half-widths of the confidence intervals were then calculated for each method and plots of the calculated values vs. the number of blocks sampled were made for each method (Fig. 6-5 and 6-6).

In addition to the controlled experiments, the two methods were also compared using data presented by Hajek and Steers (1981) from 15 randomly

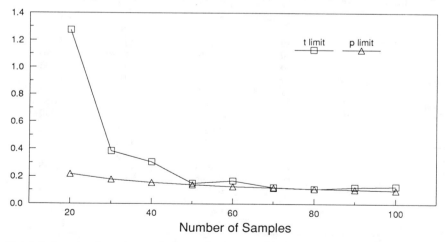

Fig. 6-3. A comparison of the half-width of the confidence intervals calculated by the *t*-statistic with those calculated by the binomial method.

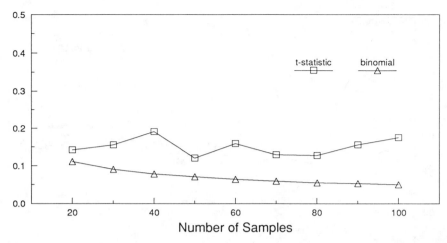

Fig. 6-4. A comparison of the estimated standard deviations from the classical method with those from the binomial method.

Table 6-3. The estimated proportion of shaded area in the blocks and the associated standard deviation based on n randomly selected samples from each of a randomly selected blocks.

a	n	an	Proportion	s_1†	s_2‡
2	5	10	0.300	0.424	0.145
3	5	15	0.467	0.115	0.129
4	5	20	0.450	0.100	0.111
5	5	25	0.520	0.268	0.100
6	5	30	0.333	0.242	0.086
7	5	35	0.514	0.227	0.084
8	5	40	0.450	0.141	0.079
9	5	45	0.400	0.200	0.073
10	5	50	0.420	0.239	0.070
2	10	20	0.550	0.354	0.111
3	10	30	0.367	0.306	0.088
4	10	40	0.425	0.150	0.078
5	10	50	0.540	0.089	0.070
6	10	60	0.483	0.160	0.065
7	10	70	0.557	0.162	0.059
8	10	80	0.563	0.106	0.055
9	10	90	0.522	0.148	0.053
10	10	100	0.470	0.189	0.050
2	15	30	0.533	0.000	0.091
3	15	45	0.578	0.039	0.074
4	15	60	0.417	0.148	0.064
5	15	75	0.480	0.137	0.058
6	15	90	0.456	0.098	0.053
7	15	105	0.543	0.112	0.049
8	15	120	0.525	0.090	0.046
9	15	135	0.489	0.088	0.043
10	15	150	0.533	0.104	0.041

† s_1 = Standard deviation from the classical method.
‡ s_2 = Standard deviation from the binomial method.

EVALUATING MAP UNIT COMPOSITION

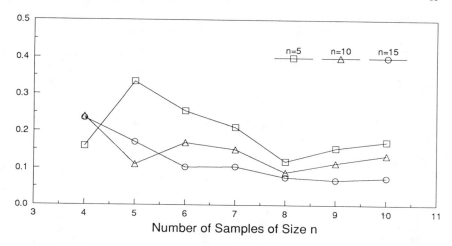

Fig. 6–5. Influence of the number of samples taken from each block on the half-width of the confidence intervals calculated by the *t*-statistic.

selected transects in Mobile County, AL. A transformed version of the original data set showing the number of points on each transect that were identified as belonging to each of the four major series located within the area as well as the total number of points on each transect is presented in Table 6–4. To compare the two analytical methods and examine the influence of the number of transects on the half-width of the confidence intervals calculated by each method, 2 through 15 transects were selected at random

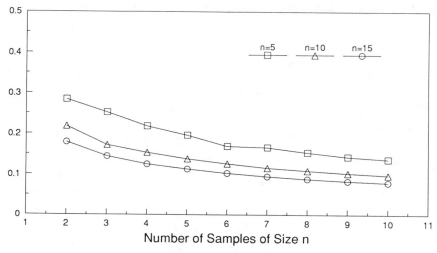

Fig. 6–6. Influence of the number of samples taken from each block on the half-widths of the confidence intervals calculated by the binomial method.

Table 6-4. Data from 15 transects in Alabama as reported by Hajek and Steers (1981). The number given is the number of points on each transect identified as the given soil series.

Transect	Soils					n
	Norfolk	Troup	Esto	Eustis	Others	
1	0	2	2	3	3	11
2	2	3	2	4	2	12
3	1	5	2	4	2	14
4	2	0	1	0	4	11
5	2	6	4	2	2	14
6	5	5	2	2	1	15
7	5	5	2	1	2	16
8	8	0	2	0	2	13
9	3	2	5	0	1	11
10	7	1	2	1	2	12
11	2	2	5	3	4	14
12	2	2	0	1	5	12
13	5	0	2	4	4	12
14	1	4	0	4	2	11
15	5	4	2	0	1	12
Total	50	41	33	29	37	190

from the 15 available transects, and the half-widths of the confidence intervals for each of the four major series were plotted against the number of transects used to calculate them (Fig. 6-7 to 6-10).

RESULTS

Results from the first experiment revealed that, although both methods gave the same estimate for the proportion of the blocks that was shaded and

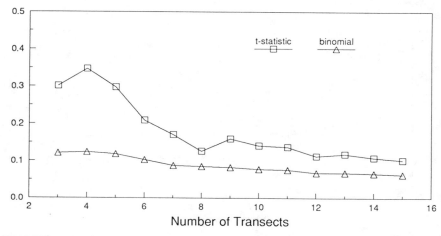

Fig. 6-7. A comparison of the half-width of the confidence intervals calculated by the t-statistic with those calculated by the binomial method for the Norfolk soil (fine-loamy, siliceous, thermic Typic Kandiudult).

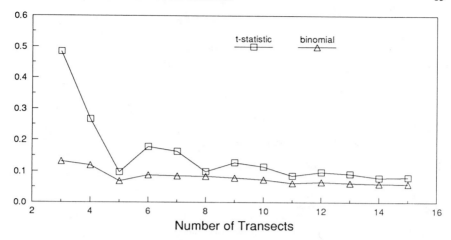

Fig. 6–8. A comparison of the half-widths of the confidence intervals calculated by the *t*-statistic with those calculated by the binomial method for the Troup soil (loamy, siliceous, thermic Grossarenic Kandiudult).

the resulting confidence intervals generally contained the true proportion; the confidence intervals calculated by the binomial method were consistently narrower than those calculated by the *t*-statistic when less than 50 samples were used (Fig. 6–2 and 6–3). A paired *t*-test that compared the half-width of the confidence intervals calculated by the *t*-statistic with those calculated by the binomial method (Table 6–2) revealed that the observed difference was significant at the 85% confidence level and that the average difference was nearly 0.17 units.

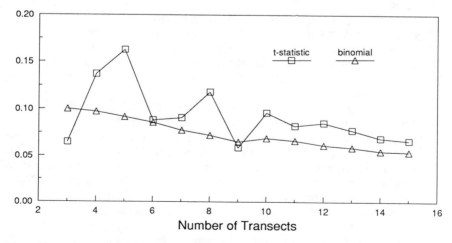

Fig. 6–9. A comparison of the half-width of the confidence intervals calculated by the *t*-statistic with those calculated by the binomial method for the Esto soil (clayey, kaolinitic, thermic Typic Kandiudult).

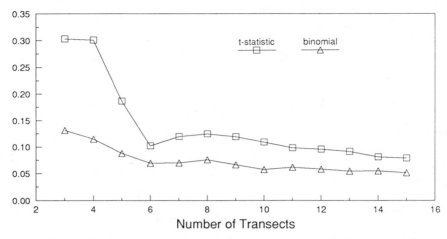

Fig. 6-10. A comparison of the half-width of the confidence intervals calculated by the *t*-statistic with those calculated by the binomial method for the Eustis soil (sandy, siliceous, thermic Psammentic Paleudult).

The half-width of the confidence interval is given by the right-hand side of Eq. [2-4]. In Eq. [2] it is the product of the Z-statistic and the estimated standard deviation for a binomial distribution; while in Eq. [4] it is the product of a *t*-statistic and the estimated standard deviation for a normal distribution divided by the square root of *n*. The value of the *t*-statistic used in Eq. [4] varied from 12.706 with one degree of freedom to 2.262 with nine degrees of freedom, while the Z-statistic used in Eq. [2] was always 1.96, which is equivalent to a *t*-statistic with infinite degrees of freedom. This difference may have contributed more to the observed differences in the half-widths of the confidence intervals than did the differences in the estimated standard deviations. To test this, the estimated standard deviations from each method were plotted vs. the total number of samples (Fig. 6-4). The plot revealed that the standard deviations used in Eq. [4] were consistently greater than those used in Eq. [2]. When they were compared by using a paired *t*-test (Table 6-2), the observed difference was found to be significant at the 99% confidence level.

The results from the second experiment revealed that, in general, as the number of samples per block increased, the half-width of the confidence intervals decreased (Fig. 6-5 and 6-6). The only exceptions were those calculated by the *t*-statistic when fewer than six blocks were sampled. These exceptions are due primarily to the large amount of variation in the widths of the confidence intervals associated with the variation in the individual block estimates of the proportion of shaded area used to obtain the overall estimate. This variation was not present in the binomial estimate since individual block estimates were not used to estimate the overall composition of the blocks.

When the two analytical methods were compared using the transect data from Alabama (Hajek & Steers, 1981), the half-widths of the confidence intervals calculated by the *t*-statistic were generally larger and more variable than those calculated by the binomial method (Fig. 6-7 to 6-10). In addition, since the number of points sampled varied from transect to transect, the estimated proportion of the survey area determined by the classical method was different from that determined by the binomial method.

DISCUSSION

Results from this study indicate that, of the two methods presently used to estimate map unit composition, the binomial method appears to give the best results, particularly when small sample sizes are used. Also, the binomial method is not influenced by block to block variation in the estimates of map unit composition and appears to be less affected by the unequal sample sizes associated with delineations of variable size. One of the primary considerations in designing a sampling scheme to estimate map unit composition is that the sampling density needs to be the same for all delineations sampled. As a result, the number of samples taken within a given delineation would be a function of the size of that delineation. Although the traditional method for estimating map unit composition as an average of the composition of the individual delineations provides similar results; its use should probably be restricted to analyzing the variability of selected soil properties unless steps are taken to remove the bias associated with delineations of different sizes.

Applications of geostatistical techniques to studies of map unit composition appear to be limited, due primarily to the large number of samples required to estimate the semivariogram and the site-specific nature of the analysis. Geostatistics can, however, be used to determine the minimum sample spacing necessary to insure that samples are nearly independent relative to selected soil properties, provided the estimates are used only for similar soils on similar geologic materials and similar landforms as those used to generate the semivariograms. More research effort is needed to determine to what extent geostatistical information from a given area can be extrapolated to similar areas.

REFERENCES

Amos, D.F., and E.P. Whiteside. 1975. Mapping accuracy of a contemporary soil survey in an urbanizing area. Soil Sci. Soc. Am. Proc. 39:937-942.

Arnold, R.W. 1979. Strategies for field resource inventories. Agronomy Mimeo 79-20. Dep. of Agron., Cornell Univ., Ithaca, NY.

Arnold, R.W. 1981. Binomial confidence limits as estimates of classification accuracy. Agronomy Mimeo 8-7. Dep. of Agron, Cornell Univ., Ithaca, NY.

Bascomb, C.L., and M.G. Jarvis. 1976. Variability in three areas of the Denchworth soil map unit. I. Purity of the map unit and property variability within it. J. Soil Sci. 27:420-437.

Beckett, P.H.T., and P.A. Burrough. 1971. The relation between cost and utility in soil survey. IV. Comparison of the utilities of soil maps produced by different survey procedures, and to different scales. J. Soil Sci. 22:467-480.

Bigler, R.J., and K.J. Liudahl. 1984. Estimating map unit composition. Soil Surv. Horiz. 25:21-25.

Brubaker,, S.C. 1989. Evaluating soil variability as related to landscape position using different statistical methods. Ph.D. diss. Texas A&M University, College Station (Diss. Abstr. 89-21686).

Burgess, T.M., and R. Webster. 1980. Optimal interpolation and isarithmic mapping of soil properties. II. Block kriging. J. Soil Sci. 31:333–341.

Burrough, P.A. 1991. Sampling designs for quantifying map unit composition. p. 89–125. *In* M.J. Mausbach and L.P. Wilding (ed.) Spatial variabilities of soils and landforms. SSSA Spec. Publ. 28. SSSA, Madison, WI.

de Gruijeter, J.J., and B.A. Marsman. 1985. Transect sampling for reliable information on mapping units. *In* D.R. Nielsen and J. Bouma (ed.) Soil spatial variability. Proc. Workshop ISSS and SSSA, Las Vegas, NV. 30 Nov.–1 Dec. 1984. PUDOC, Wageningen, Netherlands.

Edmonds, W.J., and M. Lentner. 1986. Statistical evaluation of the taxonomic composition of three soil map units in Virginia. Soil Sci. Soc. Am. J. 50:997–1001.

Gaston, L., P. Nkedi-Kizza, G. Sawka, and P.S.C. Rao. 1990. Spatial variability of morphological properties at a Florida flatwoods site. Soil Sci. Soc. Am. J. 54:527–533.

Hajek, B.F. 1982. Soil variability and quality soil surveys. A committee report. p. 17–21. *In* V.W. Carlisle and R.W. Johnson (ed.) Proc. Southern Regional Technical Work-Planning Conf. of the Natl. Coop. Soil Survey, Orlando, FL. 16–21 May 1982. USDA-SCS, Gainesville, FL.

Hajek, B.F., and C.A. Steers. 1981. Evaluation of map unit composition by the random transect method. p. 379–387. *In* Soil resource inventories and development planning. Soil Manage. Support Serv. Tech. Monogr. 1. USDA-SCS, Washington, DC.

Hopkins, D.G., J.L. Richardson, and M.D. Sweeney. 1987. Composition comparisons in sodic map unit delineations on the Dickinson Experiment Station Ranch Headquarters, North Dakota. Soil Surv. Horiz. 28:46–50.

Laslett, G.M., A.B. McBratney, P.J. Pahl, and M.F. Hutchinson. 1987. Comparison of several spatial prediction methods for soil pH. J. Soil Sci. 38:325–341.

McCormack, D.E., and L.P. Wilding. 1969. Variation of soil properties within mapping units of soils with contrasting substrata in northwestern Ohio. Soil Sci. Soc. Am. Proc. 33:587–593.

Miller, F.P. 1978. Soil surveys under pressure: The Maryland experience. J. Soil Water Conserv. 33:104–111.

Nordt, L., J.S. Jacob, and L.P. Wilding. 1991. Quantifying map unit composition for quality control in soil survey. p. 183–197. *In* M.J. Mausbach and L.P. Wilding (ed.) Spatial variabilities of soils and landforms. SSSA Spec. Publ. 28. SSSA, Madison, WI.

Oliver, M.A., and R. Webster. 1987. The elucidation of soil pattern in the Wyre Forest of the West Midlands, England. II. Spatial distribution. J. Soil Sci. 38:293–307.

Powell, J.C., and M.E. Springer. 1965. Composition and precision of classification of several mapping units of the Appling, Cecil, and Lloyd series in Walton County, Georgia. Soil Sci. Soc. Am. Proc. 29:454–458.

Snedecor, G.W., and W.G. Cochran. 1967. Statistical methods. Iowa State Univ. Press, Ames, IA.

Soil Survey Staff. 1951. Soil survey manual. USDA handbook no. 18., U.S. Gov. Print. Office, Washington, DC.

Soil Survey Staff, 1983. National soils handbook. USDA-SCS. U.S. Gov. Print. Office, Washington, DC.

Steers, C.A., and B.F. Hajek. 1979. Determination of map unit composition by a random selection of transects. Soil Sci. Soc. Am. J. 43:156–160.

Upchurch, D.R., L.P. Wilding, and J.L. Hatfield. 1988. Methods to evaluate spatial variability. p. 201–229. *In* L.R. Hosner (ed.) Reclamation of surface-mined lands. CRC Press, Inc., Boca Raton, FL.

Wang, C. 1982. Application of transect method to soil survey problems. Land Resource Research Institute Contrib. no. 82-02. LRRI Res. Branch, Agriculture Canada, Ottawa, Ontario.

West, L.T., E.D. Bearden, and D.L. Williams. 1981. Taxonomic and interpretive composition of two map units in Central Texas. p. 297. *In* Agronomy abstracts. ASA, Madison, WI.

White, E.M. 1966. Validity of the transect method for estimating compositions of soil-map areas. Soil Sci. Soc. Am. Proc. 30:129–130.

Wilding, L.P., R.B. Jones, and G.M. Schafer. 1965. Variation of soil morphological properties within Miami, Celina, and Crosby mapping units in west-central Ohio. Soil Sci. Am. Proc. 29:711–717.

Wilding, L.P., and L.R. Drees. 1983. Spatial variability and pedology. p. 83–116. *In* L.P. Wilding et al. (ed.) Pedogenesis and soil taxonomy. I. Concepts and interactions. Elsevier Sci. Publ., New York.

York, R.S., G. Uehara, and R.L. Fox. 1982. Geostatistical analysis of soil chemical properties of large land areas. II. Kriging. Soil Sci. Soc. Am. J. 46:1033–1037.

7 Sampling Designs for Quantifying Map Unit Composition

P. A. Burrough

University of Utrecht
Utrecht, the Netherlands

ABSTRACT

This chapter reviews the major options available to designers of spatial sampling systems and illustrates some of them with recent case studies carried out in Canada, the Netherlands, and Venezuela. The chapter is concerned only with those attributes of the soil that can be measured quantitatively and does not deal with attributes measured on nominal scales. There is no general optimal sampling design for quantifying map unit composition. The most appropriate sampling design (and hence numbers of samples and cost) depends on the survey aims and the spatial variation of the soil attributes studied. Different sampling approaches must be used depending on whether general or specific information is sought, and whether sampling is to determine the degree of spatial variation and to model it using the variogram before systematic sampling for mapping, or for mapping itself. The variogram must also be estimated by sampling. Sample support size in relation to survey aims and levels of local variation can be critical. Besides sample support size, the spatial resolution (scale) of the units needed to link data gathered from sampling to other spatial data is important. Alternative techniques to more intensive sampling of expensive-to-measure properties are cokriging, regression models and interactive graphics.

This chapter starts with the premise that there is no single optimal sampling design for quantifying soil map unit composition that serves all requirements. Rather, there is a range of techniques that can be used to meet different requirements and budgets depending on the type of information required, how detailed and accurate that information needs to be, and how much money is available to support the sampling program. For a given aim it should always be possible to devise a local optimum for sampling that will give the most precise and the most accurate results for a given expenditure. This optimum will not be the same in all cases. In some situations we can achieve satisfactory results with single-stage sampling of the attribute of interest; in other situations it may be necessary to adopt a stratified or even a multivariate approach. This chapter will review the major options available to designers of spatial sampling systems and will illustrate some of them with recent case studies carried out in Canada, the Netherlands and Venezuela. The chapter is concerned only with those attributes of the soil that can be measured quantitatively and does not deal with attributes measured on nominal scales.

The Aim of Sampling

The aim of sampling is to reveal information about a complex population so that meaningful statements can be made with a given degree of confidence about the whole. The concept of degree of confidence implies that the value of an attribute at an unvisited site, or the mean value of that attribute for a given block of land, can be inferred probabilistically with known variance from a sample that has been drawn without bias from the population.

The way in which the sampling problem is approached depends on what one wishes to achieve. Before sampling one must be clear on the aims. These can range from a general, qualitative description of the map unit as a population of homonymic polygons to quantitative statements about the values of single soil attributes, both individually and in conjunction with other soil attributes within the spatial extends of a single occurrence of the mapping unit.

If the aim is merely to characterize map units that have been derived from field survey or aerial photo interpretation—just to give an idea—then the conventional methods of sampling using the knowledge of skilled surveyors are probably the most cost-effective ways of working. This is providing that quantitative information is not required about the confidence limits of quantitative attributes and providing the area in question has a soil pattern that results from straightforward natural processes. When the natural patterns have been modified by human action, or are obscured by the complex interaction of spatial variation at several scales, it is not easy to deduce the way in which a given attribute varies over the landscape from the external features of the landscape alone, and designed sampling is essential.

The most effective sampling strategy for estimating the value of soil attributes in mapping units depends on many factors. If we can safely assume that the map unit can be treated as a statistically homogeneous population of individual, but unrelated entities, then we can use the methods of ordinary statistics based on normal distribution functions. In most cases, however, it is essential to examine to what degree the attributes of interest vary continuously over space and whether they are normally distributed or not. The cost of measuring an attribute and its degree of correlation with cheaper-to-measure attributes can also be important.

When a soil attribute is not distributed homogeneously over the area or volume of interest but varies continuously it is said to be *spatially dependent* or *spatially correlated*, or to exhibit *spatial variation*. The degree of spatial variation is described by the *spatial covariance structure* of the soil attribute and, as we shall see, this structure can often be modeled adequately by a function known as the *variogram*. If the soil attribute does exhibit spatial variation then the best sampling strategy (in terms of sample spacing/numbers of samples and size of samples) depends on whether we wish to predict:

1. A general mean for any given attribute in any single occurrence of the map unit.
2. A general mean for any given attribute over all occurrences of the mapping unit.
3. The spatial covariance structure of attributes within the map unit.
4. A local average for blocks of given size B and shape within the map units (i.e., to map the variation of the attribute within the delineated map unit).

In all cases the sampling program must be designed to maximize information content and to minimize costs, both in terms of field sampling and the supporting laboratory analyses.

SAMPLING WHEN THERE IS NO SPATIAL DEPENDENCE

If we can safely assume that the variation of the attribute in question is normally distributed and spatially independent (Fig. 7-1a) the number of samples needed to estimate the mean value of an attribute in a given map unit with given levels of confidence is given by conventional statistical theory. The standard formula is

$$N = t_*^2 \cdot s^2 \cdot \delta^{-2} \qquad [1]$$

where N is the number of samples, s^2 is the sample variance, and the margins of error $\pm \delta$ about the true mean * are estimated for student's t at probability level δ, usually 0.05. If the original data are nonnormally distributed, then the values of attributes of the mapping unit can be transformed by using logarithmic, logit, square root or other transformations (Webster, 1977) and Eq. [1] can be used on the transformed data.

Equation [1] implies that the number of samples required to estimate the mean within a given margin of error can be read off from nomograms like Fig. 7-2. Figure 7-2 gives the margin of error by estimating the mean in terms of the relative error percentage, that is $(\delta \mu)^{-1}$. Note that under these assumptions the geographical location of the samples is irrelevant and that data from the samples cannot be used to map local variation within the map unit.

The number of samples required to estimate the mean within given confidence limits thus depends only on the mean and variance of the whole map unit delineation. Figure 7-2 shows that if the mean is 1, then nine times as many samples are required for CL having a relative error of $\pm 90\%$ when the variance is 10 as when the variance is 1. For CL having a relative error of $\pm 40\%$, 289 samples are required instead of 25. This implies that in many situations where variances are large, the costs of sampling to estimate the mean to a required level of confidence will be prohibitive. In extreme cases,

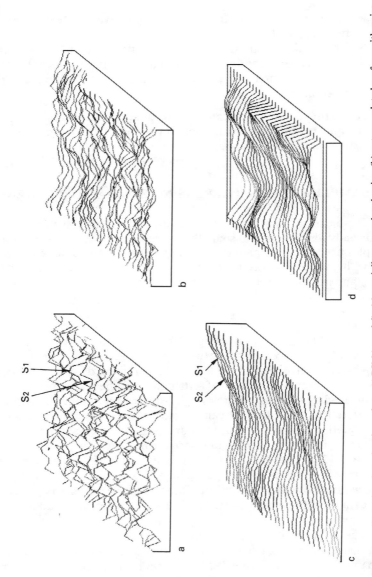

Fig. 7-1. Simulated two-dimensional stochastic surfaces (ARIMA model); (a) spatially uncorrelated noise, (b) autocorrelated surface with noise, (c) strongly autocorrelated surface, (d) surface smoothed by local moving average to simulate effects of bulking.

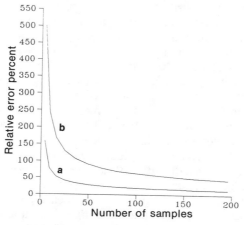

Fig. 7-2. Nomogram relating number of samples to relative error: (a) $\mu = 1$; $\sigma^2 = 1$, and (b) $\mu = 1$; $\sigma^2 = 10$.

even if costs do not play a role, taking a large number of samples could destroy the object of study!

In some situations one is not only interested in estimating mean values but also in knowing the probability that a certain value will be exceeded. If the distribution (e.g., normal or lognormal) is known then the expected frequency of extreme events can be estimated. For example with a mean of 1 and variance of 1, 5 events in every 200 should have values exceeding 3 (mean plus two standard deviations).

Estimating the Value of an Attribute at any Given Point

If the value of a soil attribute has not been measured at a given location within a map unit j, the expected value is given by the mean, μ_j. The estimation variance σ_E^2 of that estimate depends on the number of samples N_j by

$$\sigma_E^2 = \sigma_j^2 + \sigma_j^2 N_j^{-1} \qquad [2]$$

which is the sum of the within-map unit variance σ_j^2 and the estimation variance of the map unit mean (standard error squared). Clearly the estimation variance is greatly affected by the map unit variance, which as Burgess et al. (1981) point out, is why surveyors have placed so much emphasis on classification in order to make σ_j^2 as small as possible.

Choosing Sample Locations

Under conditions of spatial independence (which is often termed *statistical stationarity*) the number of samples is more important than their location or spacing. Most investigators are nevertheless unsure about the validity of this assumption and distribute their samples over the area to be investigated in order to obtain unbiased estimates of the property being sampled. There are four basic sampling strategies: complete random, fully regular (grid), stratified random, and stratified unaligned (Ripley, 1981). Each strategy can be carried out in one, two, or three dimensions. Sampling in one dimension (vertically down profiles, transects across landscape) and in two dimensions over areas are the most common.

Note that completely random sampling may produce clusters of data points that oversample some areas. When there is strong spatial dependence the clustering of samples is inefficient because each sample in the cluster will return a similar value. Regular sampling avoids this problem but can lead to bias if the sample sites coincide with a regular pattern in the landscape (e.g., drain spacing, ridge and furrow, tree spacing in orchards, etc.).

If spatial variation is of no consequence, then there are well-described standard procedures for setting up a sampling program and for estimating means and variances. For example, Gilbert (1987), Chapters 3 to 8, gives a clear exposition of the methods that can be used in this situation. But many soil attributes are spatially dependent and it is better to regard the lack of spatial variation (i.e., *spatial homogeneity*) as a special case in a broader approach to designing appropriate sampling schemes.

SAMPLING WHEN SPATIAL DEPENDENCE IS PRESENT OR IS SUSPECTED: THE PHENOMENON OF SPATIAL VARIATION

Spatial variation means that sites close together are more likely to be similar than sites spaced further apart. In other words, the values of the attribute being sampled are clustered in space which means that sampling is not as straightforward as presented above. Knowing about spatial variation can be advantageous and can lead to large savings in sampling costs and to improvements in the quality of the information gleaned from a sampling program as compared with conventional methods. The simple choropleth map model traditionally used in soil survey assumes implicitly that all spatial variation can be accounted for by the differences *between* map units. This model of homogeneous spatial units delineated by sharp boundaries is often at odds with reality. Most soil attributes vary continously over space at all spatial scales. There is no *a priori* reason to believe that the spatial variation of soil properties should become homogeneous at the level of resolution of the soil series map, or indeed at any other level of resolution—indeed, there are many good pedological and geomorphological reasons for believing otherwise.

Models of Spatial Variation

Consider Fig. 7-1a,b,c, all of which are simulations of a two-dimensional autoregressive surface having the same overall mean and variance (Schuijt, 1989). Figure 7-1a shows two-dimensional variation for which locally estimated means are very similar to the overall mean; it is statistically homogeneous or spatially independent. This type of within map unit variation is implied by the choropleth model of soil maps. Such a surface is also called "pure noise," because no information can be extracted from it beyond the parameters of mean and variance.

Figure 7-1c shows a surface in which points that are close together have a large degree of similarity and a strong spatial dependence. Put another way, the local variation of the surface is much less than the variation of the same surface over a much larger area. Obviously if such a surface were sampled properly the value of the soil attribute could be estimated more accurately at unsampled locations in Fig. 7-1c than in Fig. 7-1a. Also, a single sample (S_1) in Fig. 7-1c would be more likely to give a good idea of the value of the attribute at points in its immediate vicinity than would a single sample in Fig. 7-1a. Furthermore, the amount of extra information gained by taking a second sample (S_2) close to S_1 in Fig. 7-1c would be less than in Fig. 7-1a, because its value would be highly correlated with the first sample.

Figure 7-1b shows a situation somewhat intermediate between the others in which the longer range variation is corrupted by random variation or noise. The short range variation can be reduced in importance by bulking—that is by replacing the value at each site by an average value computed for a small, regular area (Fig. 7-1d). Bulking means that the size of the *support* increases and the level of spatial resolution decreases. Note that the support of most soil samples often is no larger than a 10-cm-diam. core while the samples are expected to give useful information about areas of land covering many tens or hundreds of meters across. The decision to bulk or to make statements about larger units of land depends on the amount of short-range noise (local variation) and on the aims of the survey. Clearly, it is useful to know how the random surface varies on average, and the relative balance between short-range noise and long-range signal before undertaking sampling.

In order to use these ideas to assist sampling, it is first necessary to assume that the spatial variation of the soil attribute under study can be modeled by a stochastic surface like the ones shown in Fig. 7-1. Attributes that vary in this way are called *regionalized variables* and they satisfy the intrinsic hypothesis (Journal & Huijbregts, 1978). They can be described using the methods of geostatistics to estimate and quantify their spatial variation.

The theory of regionalized variables is now well-known, so only the most important aspects will be given here (see Burgess & Webster, 1980a,b; Burrough, 1986; Davis, 1986; Journel & Huijbregts, 1978; Webster, 1985). Regionalized variable theory assumes that a spatial variation of any variable can be expressed as the sum of three major components. These are:

1. A structural component, associated with a constant mean value or a polynomial trend.
2. A spatially correlated random component.
3. A white noise or a residual error term that is spatially uncorrelated.

Let x be a position in one, two, or three dimensions. Then the spatial variable Z_i at x is given by

$$Z_i(x) = m(x) + \epsilon'(x) + \epsilon'' \qquad [3]$$

where $m(x)$ is a deterministic function describing the structural component of Z_i at x; $\epsilon'(x)$ is the term denoting the stochastic, locally varying spatially dependent residuals from $m(x)$; and ϵ'' is a residual, spatially independent noise term having zero mean and variance σ^2. To simplify matters, we assume here that $m(x)$ is constant. The variation of the random function $\epsilon'(x)$ over space is summarized by the semivariance γ, which for a lag (sample separation) h is given by

$$\gamma(h) = 1/2\ E(Z_{x+h} - Z_x) \qquad [4]$$

where E is the expected value. For a one-dimensional transect of n sites, the semivariance at lag h is estimated by

$$\hat{\gamma}(h) = \frac{1}{2(n-h)} \sum_{i=1}^{n-h} (Z_{xi+h} - Z_{xi})^2 \qquad [5]$$

where h is the distance between $(n - h)$ pairs of sample sites Z_i, Z_{i+h}. The function can also be estimated for sample sites in two and three dimensions and for different directions to determine possible anisotropy. A plot of $\hat{\gamma}(h)$ vs. h is called an experimental variogram. Various theoretical models [spherical, linear, exponential, gaussian, De Wijssian, Bessel functions; see Journal & Huijbregts (1978) or Webster (1985) for details] can be fitted through the experimental variogram in order to describe how the semivariance attribute values varies with sample spacing.

Features of the Variogram

Variograms may take many forms but they can be summarized in two kinds, namely nontransitional and transitional variograms.

With nontransitional variograms the semivariance increases monotonically with increasing sample spacing (Fig. 7-3a). The driving process could be a global trend or Brownian motion (fractal). If the process is fractal it means that the spatial variation is statistically "self-similar" (i.e., it shows the kind of spatial variation at whatever resolution it is studied, see Burrough [1983a]). Nontransitional variograms indicate that the variance of the property depends partly on the size of the area sampled, which is not the case for an attribute that is second-order stationary (second-order stationary spatial variation means that once local spatial variation has been neglected, the mean and variance of the attribute are independent of the size of the area studied). Nontransitional experimental variograms can be modeled by linear or De Wijssian variogram models (see Journel & Huijbregts, 1978). Equation [1] cannot be used because the variance is not constant with the size of the area.

Second-order stationary attributes can be described by transitional variogram models. With transitional variograms (Fig. 7-3b) the semivariance

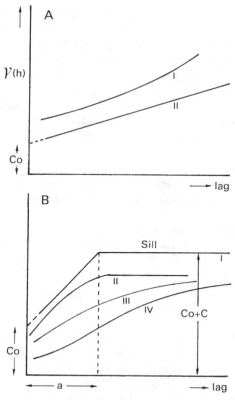

Fig. 7-3. (A) Nontransitional variograms; (i) linear, (ii) intrinsic random function. (B) Transitional variograms; (i) linear with sill, (ii) spherical, (iii) exponential, (iv) Gaussian; a = range; Co = nugget; and Co+ C = sill.

increases with sample spacing up to a critical distance called the *range* at which it levels out. The value of semivariance beyond the range is termed the *sill*. Beyond the range there is no spatial correlation between sample sites: here we are in a region that can be described by Eq. [1] because the estimated variance does not vary with sample spacing or with the size of the area, providing of course that samples are spaced further apart than the range and that the area is not so large that other sources of variation are encountered. Up to the range the semivariance increases with sample spacing as in the nontransitional case. Both kinds of variogram may show a positive intercept with the y-axis, which estimates the *nugget* variance. The *nugget* estimates the nonspatially correlated noise term ϵ'' in Eq. [3] caused by measurement errors and very short-range spatial variation below the resolution of the sampling net. Transitional variograms can be modeled by spherical, circular, exponential, Bessel functions or gaussian models.

Besides these simple models, variograms may show a variety of complex forms. Anisotropy in the underlying spatial pattern will be revealed by the experimental variograms having different slopes, ranges and possibly sills and nuggets, when estimated from pairs of samples oriented in different directions. When several spatial patterns have been superimposed upon each other (c.f., Burrough [1983a,b], Journel & Huijbregts [1978], Taylor & Burrough [1986]) the variogram will be a composite embodying variation at all the scales sampled. Periodic variation yields variograms with a periodic form. Complex experimental variograms can be modeled by sets of the variogram models given above (e.g., a double-spherical model in which the ranges of the models match two distinctly separate scales of spatial variation).

Using the Variogram to Optimize Sampling for Mapping the Relation between Sampling, Spatial Variation and Block Size

Once the variogram is known, the value of an attribute at any point in a mapping unit can be predicted from the available data points. The error of the prediction depends only on the variogram, the number and configuration of the data points and the size of the block for which the estimate is made. Knowledge of the variogram can substantially reduce the number of samples required to make predictions of mean values (or location-specific point or block estimates) for a given prediction error compared with the classical model (Eq. [1]). Note that geostatisticians speak of prediction rather than estimation because the methods involve procedures, such as the fitting of variogram models, for which it is very difficult to compute the associated degrees of freedom (Cressie [1982], and Taylor & Burrough [1986]). The following material is taken from the work of Burgess et al. (1981), McBratney and Webster (1983a), and Webster and Burgess (1984).

Using the Variogram for Interpolation

Point Estimates. The fitted variogram can be used to determine the weights λ_i needed for predicting the value of an attribute Z at any unsampled point x_0 from measurements of Z at points x_i. The estimate is a linear-weighted sum

$$\hat{Z}(x_0) = \sum_{i=1}^{n} \lambda_i \cdot Z(x_i) \qquad [6]$$

with $\sum_{i=1}^{n} \lambda_i = 1$.

The weights λ_i are chosen so that the estimate $\hat{Z}(x_0)$ is unbiased, and the prediction variance σ_E^2 is less than for any other linear combination of the observed values.

The minimum variance of $Z(x_0)$ is obtained when

$$\sum_{j=1}^{n} \lambda_j \cdot \gamma(x_i, x_j) + \varphi = \gamma(x_i, x_0) \qquad \text{for all } i, \qquad [7]$$

and is
$$\sigma_E^2 = \sum_{j=1}^{n} \lambda_j \cdot \gamma(x_j, x_0) + \varphi. \qquad [8]$$

The quantity $\gamma(x_i, x_j)$ is the semivariance of Z between the sampling points x_i and x_j; $\gamma(x_i, x_0)$ is the semivariance between the sampling point x_i and the unvisited point x_0. Both these quantities are obtained from the fitted variogram. The quantity φ is a Lagrange multiplier required for the minimalization. The prediction technique is known as *ordinary kriging*.

Block Estimates (Block Kriging). Equations [6] and [8] give estimates of the attribute and its prediction variance at unvisited sites for areas or volumes of soil that are the same size as that of the original sampling support. Very often one wishes to estimate local average values for areas that are larger than units that can be practically sampled, such as the area under an experimental plot or that covered by a remotely sensed pixel or equivalent grid cell in a raster geographical information system. This can be achieved by modifying the kriging equations to estimate an average value of Z over a block B.

The average value of Z over a block B, given by

$$Z(x_B) = \int_B \frac{Z(x)\, dx}{\text{area } B} \quad \text{is estimated by}$$

$$Z(x_B) = \sum_{i=1}^{n} \lambda_i \cdot Z(x_i) \qquad [9]$$

with $\Sigma_{i=1}^{n} \lambda_i = 1$, as before. Although Eq. [9] looks identical to Eq. [6], the weights λ_i are calculated using average semivariances between the data points and all points in the block (Burgess & Webster, 1980b).

The minimum prediction variance for the block B is now

$$\sigma_B^2 = \sum_{j=1}^{n} \lambda_j \gamma(x_j, x_B) + \varphi_B - \gamma(x_B, x_B) \qquad [10]$$

and is obtained when

$$\sum_{j=1}^{n} \lambda_j \gamma(x_i, x_B) + \varphi_B = \gamma(x_i, x_B) \text{ for all } i. \qquad [11]$$

Given the variogram (or other estimate of the covariance function) Eq. [6], [7], [9] and [10] permit the prediction of the value of an attribute Z at any location within the map unit for blocks of land having a minimum area of the sample support or larger. By predicting the value of Z at points on a regular grid the attribute can be mapped within the map unit, and a map of the associated prediction errors can be made.

How Prediction Errors can Vary with Sampling Configuration, Numbers of Data Points and other Factors

Note that because Eq. [8] and [10] show that the prediction errors of Zx_B are controlled only by the variogram and the sampling configuration, once the variogram is known it is possible to design sampling strategies that will result in any required minimum interpolation error. In particular, the prediction error σ_B^2 for a block of land B depends on the following:

1. The form of the variogram (linear, spherical or other function), the presence of anisotropy or nonnormality and the amount of nugget variance or noise.
2. The number of neighboring data points used to compute the point or block estimate.
3. Sampling, configuration—which is most efficient—irregular sampling, or a regular square grid or a triangular grid?
4. The size of the block of soil for which the estimate is made—is it an area equivalent in size to the original sample or a larger block of land?
5. How the sample points are arranged with respect to the block.

Each of these five points will be examined in turn for their effect on the prediction error. They can be evaluated using a computer program called OSSFIM (Optimal Sampling Scheme for Isarithmic Mapping)(McBratney et al., 1981; McBratney & Webster, 1981), and the following text is based on this source.

The Form of the Variogram

Data gathered during the Lacombe survey (MacMillan et al., 1987) will be used to illustrate how the form of the variogram used to compute the interpolation weights affects prediction errors. The original soil data were collected from 154 soil pits located on a 60- by 60-m grid from the Lacombe experimental farm, Alberta, Canada. The data used here are the percentage sand content of the A horizon (ASAND) and the Na content of the C horizon (CNa).

Figure 7-4a shows the experimental variogram for ASAND that was fitted by an isotropic spherical model

$$\gamma(h) = 159.6 + 186.6 \, [3/2 \, (h/153.4) - 1/2 \, (h/153.4)^3].$$

The CNa has a weaker spatial dependence as shown by the large nugget variance. The maximum kriging standard errors are given in Table 7-1b. In order to relate the two sets of results better, Table 7-2 presents the same data for both properties converted to relative error percentage in terms of the overall means and variances.

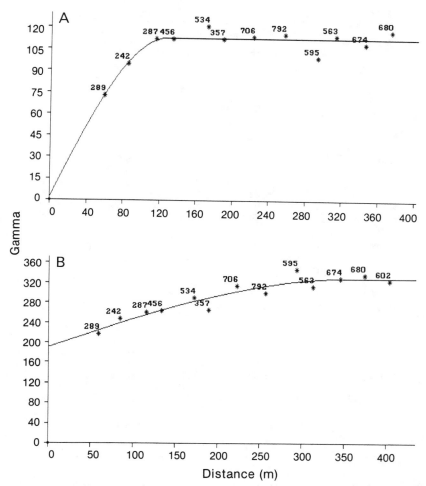

Fig. 7-4. (A) Experimental variogram for Lacombe ASAND percentage fitted by spherical model. (B) Idem CNa content cmol kg^{-1}. The numbers of pairs of samples are given for each estimate of the semivariance.

The results in Tables 7-1a,b, 7-2a,b show the value of the variogram in reducing the maximum standard error of point estimates when spatial dependence is strong, as is the case with the ASAND data. The improvements over the estimated population standard deviation are particularly marked when the sample spacing is less than the range.

When the spatial dependence is weaker, as is the case with the CNa data, then the improvements are less marked, but are still worthwhile. In all cases the kriging estimates of relative error are considerably lower than the classical estimates, even when samples are located 100 m apart.

The nugget limits the minimum value of the maximum prediction variance and so fitting a variogram model with a smaller nugget will reduce the

Table 7-1. Maximum kriging standard errors at points within the mapping unit.

Sample spacing	Kriging standard error (percentage sand)			
	Number of points			
	25	16	9	4
m				
	A. Lacombe ASAND percentage			
10.0	3.05	3.05	3.05	3.06
20.0	4.06	4.06	4.07	4.09
30.0	4.88	4.88	4.89	4.92
40.0	5.57	5.58	5.61	5.65
50.0	6.19	6.21	6.24	6.33
60.0	6.77	6.80	6.84	6.96
70.0	7.33	7.36	7.43	7.58
80.0	8.03	8.04	8.09	8.18
90.0	8.74	8.75	8.76	8.78
100.0	9.33	9.34	9.34	9.35
Classical estimate Overall mean = 26.64 SD = 10.8	11.01	11.13	11.38	12.07
	B. Lacombe CNa percentage			
10.0	13.40	13.45	13.73	14.36
20.0	13.80	13.82	14.08	14.60
30.0	14.15	14.16	14.38	14.83
40.0	14.45	14.50	14.65	15.05
50.0	14.73	14.74	14.91	15.27
60.0	14.99	14.99	15.14	15.48
70.0	15.23	15.23	15.38	15.69
80.0	15.46	15.46	15.59	15.89
90.0	15.67	15.67	15.80	16.08
100.0	15.88	15.87	15.99	16.27
Classical estimate Overall mean = 14.89 SD = 18.2	18.56	18.76	19.18	20.35

prediction variances. The errors associated with estimating and fitting variogram models are discussed in "Estimating the Variogram." The nugget can also be reduced by increasing the support size or by bulking, which will reduce the maximum prediction error.

The Number of Neighbors Used in the Computations

The data in Tables 7-1 and 7-2 also illustrate how the number of data points used to compute predicted values at unsampled points affects prediction errors. In the situation here where the data points are on a regular grid, using more near neighbors to compute the predicted value for ASAND yields little improvement because of the strong spatial covariance. Using more near neighbors to predict values of CNa yields relatively larger improvements because spatial dependence is not so strong. In both cases using 10 nearer neighbors seems a reasonable compromise between accuracy and computation times.

QUANTIFYING MAP UNIT COMPOSITION

Table 7-2. Kriging relative errors and classical relative errors for estimates at points within the mapping unit.

Sample spacing	Kriging relative error (percentage)			
	Number of points			
	25	16	9	4
m				

A. Maximum kriging relative errors (referred to a mean of 26.64%) for Lacombe ASAND as a function of sample spacing and numbers of points used to compute the predicted value.

10.0	11.5	11.5	11.5	11.5
20.0	15.2	15.2	15.3	15.4
30.0	18.3	18.3	18.4	18.5
40.0	20.9	21.0	21.1	21.2
50.0	23.2	23.3	23.4	23.8
60.0	25.4	25.5	25.7	26.1
70.0	27.5	27.6	27.9	28.5
80.0	30.1	30.2	30.4	30.7
90.0	32.8	32.9	32.9	33.0
100.0	35.0	35.1	35.1	35.1
Classical estimate (Eq. [2])	41.0	42.0	43.0	45.0

B. Maximum kriging relative errors for Lacombe CNa (referred to a mean of 14.89 cmol kg^{-1}) as a function of sample spacing and numbers of points used to compute the predicted value.

10.0	90	90	92	96
20.0	91	93	95	98
30.0	95	95	97	100
40.0	97	97	98	101
50.0	99	99	100	103
60.0	101	101	102	104
70.0	102	102	103	105
80.0	104	104	105	107
90.0	105	105	106	108
100.0	107	107	107	109
Classical estimate (Eq. [2])	125	126	129	137

The Effect of Sample Configuration

Unless the variogram is 100% nugget (no spatial dependence), the closer the data points to the point to be estimated, the better that estimate will be. The maximum distance between sample points and unsampled points depends on the sample configuration, which could be (i) an equilateral triangular grid, (ii) a square grid, (iii) from irregularly (randomly) distributed points.

A. **Triangular Grid.** For equilateral triangular cells of unit area each having one observation located at the center, the sample spacing is 1.0746 units (triangle side) and the maximum distance between a data point and an unsampled point is 0.6204 units.

B. **Square Grid.** For unit cells of side 1.00 units and one observation per cell the maximum distance between a data point and an unsampled point is 0.7071 units.

C. **Irregular Sample.** Although the minimum sampling distance may be less than that on unit grids, the average minimum distance between a sample site and an unsampled point will be larger than for the grids for the same number of data points.

Conclusion. Sampling on an equilateral triangular grid will yield the smallest errors for predictions of attribute values at unsampled points. Square grids are nearly as efficient and have the additional advantage that they are easier to lay out in the field and to handle in the computer. Irregularly spaced data are the least efficient for mapping. When anisotropy is present the square grids can be elongated to rectangles to achieve optimal results in all directions. In the rest of this chapter all statements about optimal sample spacing and sampling density will refer to a square grid.

Computing Estimates for Blocks of Land that are Larger than the Support Size (Block Kriging)

Block kriging reduces prediction errors because the predicted attribute values and their standard errors refer to larger blocks of land than the original samples. The previous arguments on reducing prediction error also apply to block kriging but some modifications are necessary because of complications caused by the relation between block size and sample grid size, and the location of a block relative to the sampling grid. These relations affect the values of the average semivariances between the data points and the block that are used to compute the interpolation weights (see Eq. [9]).

When the block is small in relation to grid size, its maximum prediction error lies between two extremes, depending on whether the block is situated over a data point, or if it is located at the center of a grid square at a maximum distance from all four corner data points. In the first case the maximum prediction error will be low as the single data point contributes greatly to the block estimate; in the second case the four data points at the corner of the cell contribute equally to the prediction at the center point but because they are as far away as possible the prediction error is at a maximum. As the block size increases these two situations become less extreme. Blocks centered over sample points receive a decreasing contribution from that observation as they increase in size and so the maximum prediction variance increases. This continues until the block is large enough to be influenced by the next nearest neighbor data points on the grid when the maximum prediction variance decreases. Increasing the block size thereafter causes the prediction variance to increase again.

For blocks located at the centers of data cells the prediction variance decreases as the block size increases because the data points on the grid are closer to the block edges and therefore the weight of their contribution increases. As block size increases there will come a point when the prediction variance for grid point-centered blocks exceeds that of grid cell-centered blocks. The minimum value of the prediction variance achievable with block kriging for a given block size when using data from a regular grid is thus given by the maximum value of these two extremes.

QUANTIFYING MAP UNIT COMPOSITION

Knowledge about how the maximum prediction error varies with block size can be used to design optimal sampling strategies. The maximum prediction error resulting from any combination of regular sample spacing and block size can be calculated for any variogram to yield a prediction error surface as shown in Fig. 7-5. This figure (which was computed with the OSSFIM program, McBratney & Webster, 1981) shows the prediction errors that can be expected from an existing survey on a regular grid, but it also allows the best combination of block size and sampling interval to be chosen in order to achieve specified minimum prediction errors, before sampling is undertaken. Therefore grid surveys can be tailored to yield results with a required prediction error for a given level of spatial resolution (block size).

The Lacombe data provide a convenient example. Figure 7-5 shows a contour plot of maximum kriging standard error for the Lacombe ASAND data for various combinations of block size (10–100 m) and sample spacing (10–100 m). The figure shows that if sample spacing were increased to save money (a coarser sampling grid) then a specified level of the prediction error of a block mean could only be obtained by sacrificing spatial resolution. However, if both resolution and standard error are specified, diagrams such as Fig. 7-5 allow the required sample spacing to be determined. For exam-

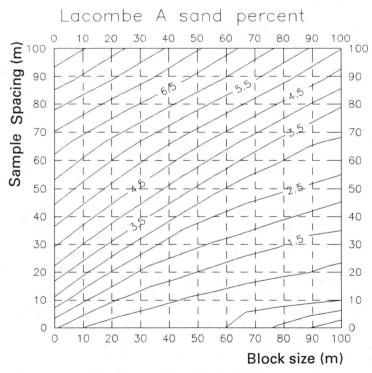

Fig. 7-5. Isolines of maximum block kriging prediction error for both cell-centered and grid-centered blocks for combinations of sample spacing (y axis) and block size (x axis) for Lacombe ASAND percentage.

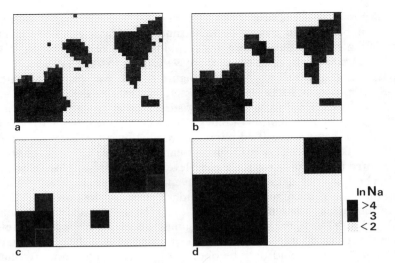

Fig. 7-6. Maps of Lacombe CNa content having equal maximum block kriging prediction error of 6 cmol$_c$ kg^{-1} derived from different combinations of sample spacing and block size. (a) 20-m blocks, (b) 40-m blocks, (c) 100-m blocks, (d) 200-m blocks. Block means are given as integer values of logarithm Na(cmol$_c$ kg^{-1}).

ple, if the survey must have a resolution of 30-m blocks to match the spatial resolution of a Landsat TM satellite, and the standard error of ASAND for a grid cell mean should not exceed 5% (a 20% relative error when mean sand content is 25%) then samples can be spaced on a 60-m grid. Figure 7-6 shows how the spatial resolution of the survey must be degraded (larger blocks) if prediction errors for CNa are to remain constant at 6 cmol kg^{-1} as sample spacing increases.

Optimal Estimates for a Single Block of a Given Size —How Standard Errors may Vary with the Arrangement of the Data Points within the Block

The previous section described how maximum prediction variances can be estimated when average values are estimated for blocks of a given size over the whole extent of the mapping unit. An alternative, but common situation is the case of estimating the mean of a soil property for a single block of land within a mapping unit under the constraints of a limited sampling budget. The problem is how to locate the permitted number of samples over the block in order to obtain the best estimate of the mean. Here, too, knowledge of the variogram of the attribute to be sampled can yield considerable savings.

Using the classical technique, the location of the data points is deemed to be unimportant, though we may spread them out over the area. If we have a variogram we can locate our data points in several ways, and even if they are located on a grid there will be an optimal way to lay them out over the area.

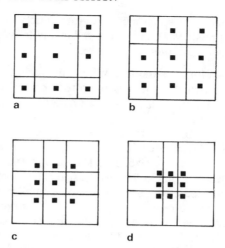

Fig. 7-7. Four different ways of arranging nine points on a square grid in a square block of land.

Figure 7-7 shows the different ways in which N data points ($N = 9$) can be laid out within a square block on a centrally located square grid. The prediction variance of a block B is the expected squared difference between the kriging prediction of the block mean and the true value. The standard error of the block mean is the square root of the prediction variance of the block B. The prediction variance is given by

$$\sigma_E^2 (B) = E \{[Z(B) - \hat{Z}(B)]^2\}$$

$$= 2 \sum_{i=1}^{n} \lambda_i \gamma (x_i, B) - \sum_{i=1}^{n} \sum_{j=1}^{n} \lambda_i \lambda_j \gamma (x_i, x_j) - \gamma(B,B) \quad [12]$$

where $\gamma(x_i,x_j)$ is the semivariance of the property between points x_i, x_j, taking account of the distance $x_i - x_j$ between them (and the angle in cases of anisotropy), *$\gamma(x_i,B)$ is the average of the semivariance between x_i and all points within the block,* and $\gamma(B,B)$ is the average semivariance within the block [i.e., the within-block variance, $\gamma(B,B)$, see Webster & Burgess, 1984].

Equation [12] shows that the prediction variances are not constant, but depend on the size of Block B, the form of the variogram and the distance between the data points (i.e., the configuration of sampling points in relation to the block to be estimated). Note that these variances do not depend on the observed values themselves (except through the variogram).

Webster & Burgess (1984) have shown that when the variogram is linear the minimum prediction variance occurs when data points are located in the middle of a set of equal Dirichlet tiles (Fig. 7-7b). For a linear variogram the efficiency of using the variogram in terms of the maximum standard error depends on the ratio r of the nugget variance to the gradient. When $r = 0$, the nugget variance is zero and the geostatistical estimate achieves maximum efficiency compared with classical estimates, which occurs when

$r = \infty$. The estimates for spherical variogram functions are more complex (see Webster & Burgess, 1984), but yield similar results. Webster and Nortcliff (1984) present other examples.

Conclusions—Block Estimates. When the variogram is a monotonic increasing function (which it usually is up to the range) kriging prediction variances for a block tend to increase the further the block is from the data points. This means that kriging prediction variances tend to increase with sample spacing simply because the distance between the data points and the block also increases. Also, because the average semivariance *within* a block tends to increase with block size, Eq. [12] shows that kriging prediction variances will *decrease* as blocks increase in size. This is intuitively correct—it is more difficult to predict the mean value of a small block to a given level of precision than it is to predict the mean level of a larger block to the same level of precision. So we arrive at the following situation. A given level of prediction variance can be achieved for a variety of combinations of block size and sample spacing. If you require fine spatial resolution (small blocks), then more dense sampling is needed to predict block means to a given level of precision, than if coarser resolution (large blocks) is acceptable. An optimal sampling scheme will attempt to find the combination of block size and sample spacing that for a given level of precision results in the lowest total costs.

THE EFFECT OF SPATIAL VARIATION ON THE CONCEPT OF A MAP UNIT AS A POPULATION OF INDIVIDUAL SAMPLES: DEFINING THE SUPPORT SIZE WHEN MAP UNITS ARE NOT SPATIALLY HOMOGENEOUS

One of the first problems encountered when sampling soil under conditions of spatial variation over many scales is the concept of the *sample unit* as a member of the population. Is an area of Soil Series X as delineated on a map to be regarded as a collection of separate soil individuals $m \epsilon P$? If so, how is the individual to be defined? Is a soil individual m equivalent in size to a spatula measurement of soil particles submitted for chemical analysis in an autoanalyzer? Is a soil individual m a 10-cm diam. and 20-cm-long soil core taken by auger? Is a soil individual m a complete soil profile pit (which implies that it is a large hole in the ground of some 2 m^3)? Is a soil individual a truck load of material that is to be added or removed from an area when tackling pollution problems? Is a soil individual m the 1-m-thick layer of soil under an experimental plot? Is a soil individual the 1- to 2-m-thick layer covered by a pixel on a remotely sensed image? Clearly, it is absolutely essential to define the size units (the *support*) that are appropriate for the soil attribute and the aim of the survey in order to make meaningful statements about the phenomenon being studied at a correct level of spatial resolution.

Results from a study by Ten Berge et al. (1983) illustrate the problems that can arise when the support size has been chosen injudiciously. The aim of the study was to characterize the spatial variation of soil surface temperature across a 350-m-wide experimental field at the Ir.A.P. Minderhoudhoeve Experimental Farm in the Flevopolders, the Netherlands. The area is extremely flat, but because of differences in the patterns of sedimentation over the farm, annual thermal infrared surveys using airborne scanners carried out in early spring in the years 1980 to 1982 detected effective surface temperature differences across ploughed experimental fields of some 0.6 °C. The aim of the study reported by Ten Berge et al. (1983) was to determine which surface properties of the soil were well-correlated with these differences over the fields, and ultimately to determine if the surface temperature variations had implications for possible crop response differences.

The thermal infrared imagery of 26 Mar. 1986 was obtained using a Daedalus DS126° Multispectral scanner (Daedalus Enterprises, Ann Arbor, MI), with a spectral window of 8 to 14 μm. The data were collected in the middle of the day and the height of the aircraft carrying the scanner was such that the pixel size of the imagery was 1.5 × 1.5 m. This means that all variations within a block of this size were averaged out to yield a single data value (a weighted average) for this support size.

At the same time that the aircraft carrying the scanner was flying overhead, data were collected on the ground. These data included the thermal infrared emission of the soil surface and information about the texture and moisture content of the 0- to 5-mm, 5- to 15-mm and 15- to 30-mm depths of the soil. The field data were collected at points spaced 4 m apart along two 200-m-long transects (50 observation points) located perpendicular to the main temperature gradients located on previous imagery. These transects were clearly marked so that they could easily be located on the thermal imagery. The thermal imagery emission data for the soil surface were collected by a portable Heimann KT-15 radiometer (Heimann Instruments), mounted on a tripod with a spatial resolution of 0.03 m^2 at the same 8- to 14-μm spectral window (approximately 0.17 by 0.17 m). The data for determining soil texture were collected by bulking six subsamples taken within an iron grid measuring 1.00 by 1.00 m laid over the soil at each sample site. Soil moisture tension was measured using a needle porous cup tensiometer 2 cm in diameter connected to a pressure transducer with digital readout. Each measurement of soil moisture tension was an average of two readings spaced 20 cm apart with the tensiometer being inserted at a sampling depth of 5 cm for an equalization time of 90 s. In addition, surface soil temperatures for the 0- to 5-cm layer were measured with a needle thermocouple with digital readout.

To sum up, the survey provided data for two 50-point transects but the data had been collected using different support sizes. The airborne infrared scanner data had a resolution of 1.5 by 1.5 m; the Heimann infrared data had a resolution of 0.17 by 0.17 m; the soil moisture data had a resolution of 0.02 m; and the soil texture data had an effective support size of 1.00 by 1.00 m.

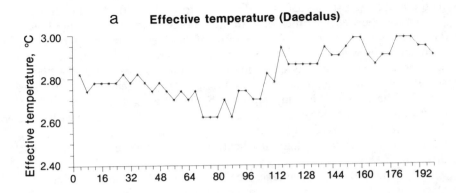
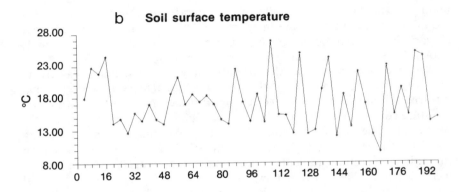
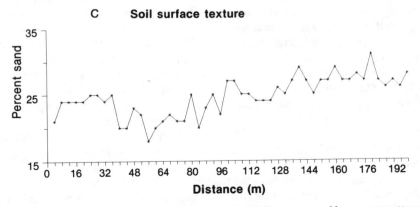

Fig. 7-8. The effect of support size and relative weights of short-range and long-range patterns on the form of the variograms as estimated for soil surfaces at the Minderhoudhoeve in 1982. (a) Airborne Daedalus thermal infrared scanner, window 150 by 150 cm. (b) Handheld Heimann scanner, window 17 by 17 cm, (c) percentage sand top 2-mm soil surface bulked within window of 100 by 100 cm.

QUANTIFYING MAP UNIT COMPOSITION

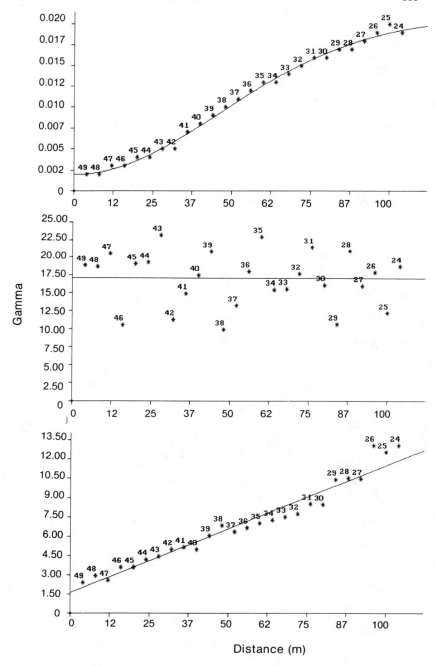

Fig. 7-8. Variograms, continued.

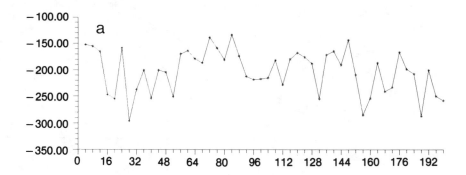
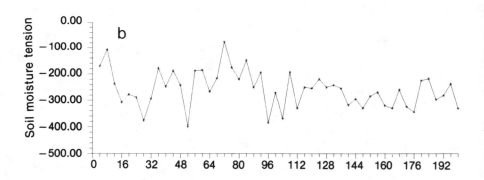
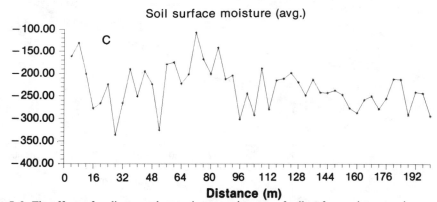

Fig. 7-9. The effects of replicates and averaging on variograms of soil surface moisture tension determined by porous cup and transducer with effective window of 1 by 1 cm; (a) Transect 1, (b) Transect 2 located 20 cm from Transect 1, (c) transect obtained by averaging transect values from Transects 1 and 2 to remove short-range variation.

QUANTIFYING MAP UNIT COMPOSITION

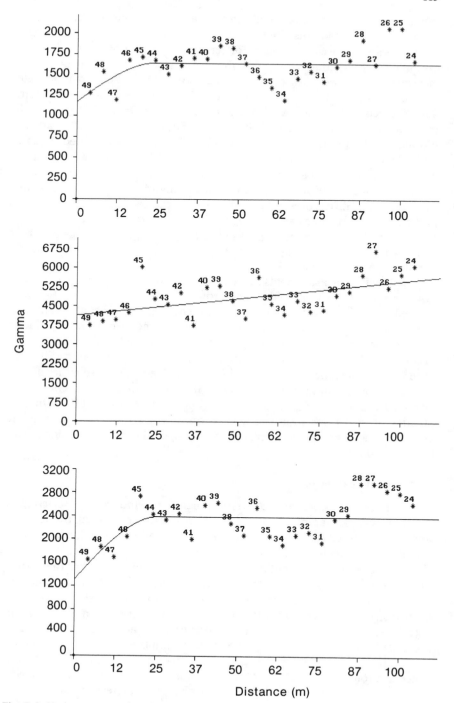

Fig. 7-9. Variograms, continued.

All data were first examined for normality, and none were found to be seriously nonnormally distributed. The effects of the different support sizes became evident however, when the different data were plotted and they were examined for spatial variation. In particular the two infrared data measurements showed little relation to each other. The airborne data reflected the well-known coherent spatial variation over the field but the Heimann data showed only short-range variation or noise. Observations with the thermocouple soon showed why this was so. The experimental field had been plowed and had plow ridges of some 15 to 20 cm high separated by 25 to 30 cm. One side of the plow ridges was warmed by the low March sun to yield surface soil temperatures of 20 to 21 °C; the other side of the plow ridges were in shadow and temperatures were only 14 to 15 °C. Consequently the short-range differences in temperature were causing the large fluctuations in the Heimann readings and swamped any weak variations across the field. Because the Daedalus scanner pixels were large enough to average out these short-range variations they returned information about the longer-range variations over the field.

Soil texture measurements made on bulked soil samples made up of nine randomly located subsamples taken from support areas of 1 by 1 m also showed a recognizable degree of correspondence with the Daedalus data (Fig. 7-8). These soil texture data also yield a smooth monotonic variogram. In contrast, measurements of soil moisture tension made in duplicate over distances of 20 cm that were made using the portable soil tensiometer showed little spatial coherence over the field (Fig. 7-9) when analyzed either as two separate transects or as a single transect of local average values. Only when all data are analyzed related to their location can it be seen that the sampling practice of measuring moisture tension on top of a plow ridge and at the bottom (to get a good local average) has swamped the long-range signal with local periodic noise (Fig. 7-10).

Summary: Creating the Correct Support Size

1. Extreme short-range variation can be removed by selecting a sufficiently large sample to study the attribute of interest at the level of resolution required for the purposes of the investigation. Sampling large areas is really only feasible with remote sensing.

2. Subsampling and bulking within an area/volume that has the correct support size can achieve a similar result to sampling a larger area completely. This may require some information or assumptions about the distribution of the attribute values within the basic sampling unit (e.g., normality) and it requires that the attributes being measured are additive. For example, because pH is measured on a logarithmic scale simple averages of pH readings cannot be computed and the data must first be transformed to H ion concentrations. Subsamples for bulking should not be located in such a way that they introduce effects of periodicity or regular differences at the local scale.

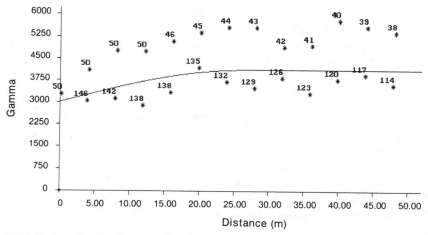

Fig. 7-10. As in Fig. 7-9, but now using all data. The experimental variogram has been computed using the locational data that pairs of points were located 20 cm apart. The strong short-range periodicity is caused by the samples being located alternately on top of and between plow ridges.

ESTIMATING THE VARIOGRAM

"The most serious obstacle to using optimal sampling strategies . . . is the need to know the variogram in advance" (Webster & Burgess, 1984). It is also important to consider whether a simple variogram, and the assumption of the intrinsic hypothesis, is always appropriate, or whether more advanced generalized covariance functions should be used. Here we assume that the variogram is sufficient, but that need not always be the case (c.f., Van Geer, 1987). As shown in previous sections, the variogram has to be known in order to optimize sampling, but because it must itself be obtained by sampling it is sensible to consider how much sampling effort establishing the variogram may require. Knowledge of the sampling effort required to establish the variogram could allow rational decisions to be made about splitting sampling effort between that required for establishing the variogram and that required for mapping or for establishing local or general means. In practice it is common that data sets collected ad hoc are used for both functions. In future perhaps it will be as common for variogram functions to be recorded in soil information systems as profile descriptions and laboratory data are stored today.

The spatial covariance function can be estimated from data collected along one-dimensional transects, two-dimensional grids, two-dimensional irregularly spaced data, and nested sampling schemes.

One-Dimensional Transects can be located in various directions, giving an idea of spatial anisotropy. They are easy to lay out in the field but are limited in resolution to the basic unit interval. Because semivariances are usually estimated from one-dimensional transect data using Eq. [5] the confidence limits for the estimated $\hat{\gamma}(h)$ at each h are not simply related to the number of pairs N-h. Following Welch (1937) and Cressie (1982), Taylor and Burrough (1986) estimated the effective degrees of freedom for each lag interval h in one dimension as

$$\hat{m}(h) = 2 E [\hat{\gamma}(h)]^2 / \text{var}[\hat{\gamma}(h)] \qquad [8]$$

The variation of $\hat{m}$ with N-h depends on the relative balance of short- and long-range variation. When long-range variation dominates, then $\hat{m}$ falls off more rapidly than when short-range variation dominates (Fig. 7–11).

Simulation studies suggest that variogram estimates are more susceptible to sampling variation than one might expect. Figure 7–12 shows the results of simulating a one-dimensional transect using the nested model (Burrough, 1983b) and calculating the variogram for runs of 200, 400 and 600 points.

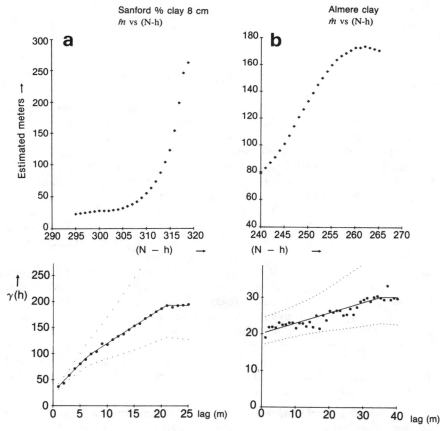

Fig. 7–11. Figure taken from Taylor and Burrough (1986) showing how the effective degrees of freedom $\hat{m}$ can vary with the number of pairs N-h used to estimate the variogram: (a) long-range variation (small nugget); (b) short-range variation (large nugget relative to sill).

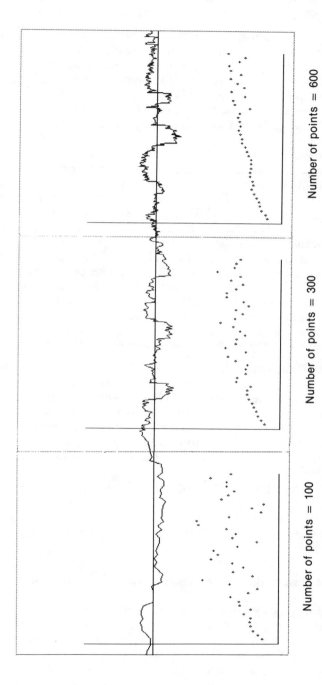

Fig. 7-12. One-dimensional simulations showing the effect of numbers of samples on the smoothness of the estimated variogram when the generating model is kept the same.

Clearly the degree of spreading in the estimates of the sill are much greater for the 200 points than for the 600 points.

Two-Dimensional Grid Surveys. When data have been collected on a regular grid laid over the mapping unit, the variogram can be estimated easily over several different directions, most notably along the axes of the grid. Compared with irregularly spaced samples (see next section), both the lag distances and the directions are known exactly. Also, the number of pairs per lag interval is a simple function of the grid spacing, and numbers of grid pairs tend not to fall off too much as lags increase, providing semivariances are not estimated for lags greater than half the maximum dimension of the area sampled. The main disadvantage is that the maximum resolution (smallest lag) is limited to the grid spacing, and it is often recommended that the basic grid be supplemented by closely spaced pairs of points at some grid nodes so that the form of the variogram at small lag intervals can be estimated better. There is no clear advice on how many samples are required; but it seems obvious that areas with clear, dominant long-range variation need fewer samples to establish the variogram than areas in which short-range variation is paramount. Using simulation, Oliver and Webster (personal communication, 1989) have experimented with estimating variograms for repeated two-dimensional simulations of the same model, and these results tend to confirm those from Taylor and Burrough (1986) and the above-mentioned one-dimensional simulations. So far there is little work published in soil science on how critical the estimates of the variograms are for determining the prediction variances and for optimizing sampling.

Two-Dimensional Irregular Data. Variograms can also be computed from irregularly spaced data, for example data collected from completely random or stratified random sampling schemes. The main disadvantage is that unlike systematic samples the pairs of points needed to compute the semivariance at different lags do not lie at discrete distances from each other. So a search window is used to find those points that are approximately at the correct distance from the first. The larger the window, the more points will be found, but less precise will be the value obtained for the lag. Figures 7–13a,b,c show how the form of the experimental variogram may vary with the size of the window. In particular, Fig. 7–13c shows that if the window is too large, then all information about short-range variation can be smoothed away leaving a variogram that only reflects the long-range components of the variation.

General Observations on Estimating Variograms and Fitting Models

All methods given above suffer from the problem of *effective degrees of freedom* noted for the one-dimensional case. Consequently fitting a variogram model should never be done using the raw bulked estimates but needs to take the effective numbers of data pairs under each estimate into account, e.g., by using methods of weighted least squares. Data need to be normally distributed before computing the sample variogram. Variograms computed

QUANTIFYING MAP UNIT COMPOSITION

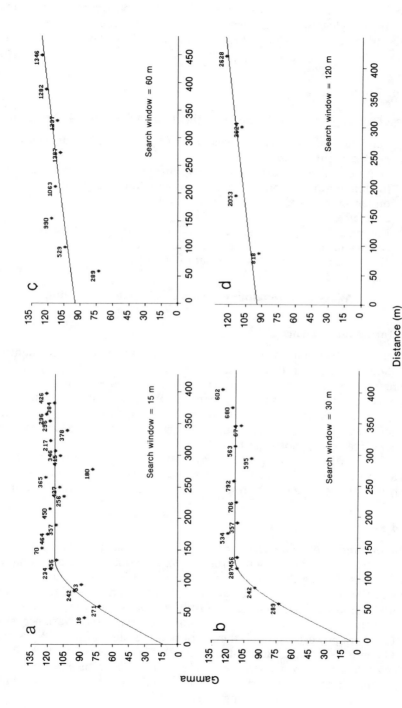

Fig. 7-13. Variation of the form of the experimental variogram and the best-fit spherical model with window size used to find pairs of data points. This example uses the ASAND data from Lacombe: (a) 15-m window, (b) 30-m window, (c) 60-m window, (d) 120-m window.

from nonnormal data will overestimate true prediction errors. Transform the data by using logarithms or other appropriate function, or by using robust statistics (Cressie & Hawkins, 1980; Rousseeuw & Leroy, 1987) when necessary.

Rules of Thumb

The following rules of thumb for estimating variograms can be given:

1. Make sure that there are sufficient pairs of points at all distance classes needed for the study.
2. Don't estimate the variogram for lags that exceed more than one-half the longest dimension of the area, because longer lags will be based on few points located at the extremes of the area and may suffer from edge effects.
3. For irregularly spaced data, choose window sizes as small as possible so that local variation is not smoothed away.
4. Don't be afraid to transform data if necessary.

Quick Ways of Estimating the Degree of Spatial Variation

Soil Information Systems

Many soil information systems now contain huge amounts of quantitative data about soil properties. If these data are linked to geographic coordinates then it is a simple matter to retrieve data for any single or multiple occurrence of a mapping unit. The programs for computing variograms are now available for personal computers for very modest sums (e.g., Rijksuniversiteit Utrecht, 1988; Englund & Sparks, 1988) so lack of computing tools is no reason for not beginning.

When insufficient existing data are available to estimate variograms then there are several ways in which variograms can be estimated quickly and cheaply though there is a need to research just how reliable these methods are.

Nested Sampling

Before undertaking any form of systematic sample it may be advantageous to perform a reconnaissance study in order to find out the best sample spacing for estimating the covariance structure and for future quantitative mapping. Short transects can be used for obtaining an idea of the degree of spatial variation (Marsman & De Gruijter, 1984), but the methods of balanced and unbalanced nested sampling and analysis of variance are probably more powerful though they require more observations (Webster, 1977; Oliver & Webster, 1986; Mateos et al., 1987; Riezebos, 1989). Nested sampling schemes allow a wider range of spatial scales to be examined for a given number of samples than with long, regularly spaced transects but they are not good for investigating anisotropy.

QUANTIFYING MAP UNIT COMPOSITION

Table 7-3. Observed cost efficiencies using nested sampling to determine spatial variation before systematic mapping in Turen, Venezuela.

A. Actual survey cost using two-stage nested approach.

The total time actually spent was
Setting up 100-m grid (4 × 9 points)   11 h × 4 p = 44 working h
Nested soil sampling (56 observations)   11 h × 6 p = 66 working h
Systematic sampling (69 observations)   = 93 working h
Total time of sampling carried out   = 203 working h
Total number of samples (3 × 69)   = 207 samples

B. Computed survey cost for a hypothetical one-stage survey

The time required for a sampling of the same field at the original "eyeball" estimate of a 50-m grid interval was estimated by

Setting up 100-m grid (4 × 9 points)   11 h × 4 p = 44 working h
Sampling 400 × 850-m: 162 observations ×   = 219 working h
  (93 h/69 observations)
Total time of sampling   = 263 working h
Total number of samples (3 × 162)   = 486 samples

The advantages of nested sampling for estimating spatial covariance and for reducing total survey costs can be seen from a survey carried out at a 500-ha experimental farm in Venezuela (Mateos et al., 1987). Eyeball estimates of the spatial variation of the young alluvial soil suggested mapping the farm on a regular 50- by 50-m grid, a total of 162 profiles. A reconnaissance nested sampling based on 56 profiles with four levels of clustered sampling at 250-m, 60-m, 10-m, and 2-m spacing was used. At each site the soil profile was sampled by hand. The field data were analyzed using a balanced nested analysis of variance (Webster, 1977) to estimate the amount of variation accruing for each extra spacing. The resulting crude variograms showed that for many soil properties large increases in variance only began to accrue when sampling intervals exceeded 60 to 80 m. These results suggested that subsequent systematic mapping could safely be based on a 75- by 75-m grid, which brought large savings both in field work and in the laboratory analyses that were used in the systematic mapping (Table 7-3). The two-stage nested sampling followed by the systematic sampling resulted in a saving of 60 working hours of field work and a reduction in the number of samples for laboratory analysis from 486 to 207.

THE PROBLEM WHEN AN ATTRIBUTE IS TOO EXPENSIVE TO MEASURE FREQUENTLY

When a soil attribute is too expensive to measure at many sites there are several ways in which the data can yield more information than a straightforward statistical or geostatistical method allows. These are (i) the methods of dynamic graphics, (ii) cokriging, and (iii) the use of regression models or other models to compute the values of expensive-to-measure attributes from cheaper data (simple transfer models, complex simulation-crop growth).

Dynamic Graphics. These are new interactive computer graphics techniques that allow complex data to be represented simultaneously in different windows on the computer screen in several ways. For example, data about the mineral content of sites along stream beds can be displayed as sets of histograms for each element analyzed, and at the same time two-dimensional or even quasi-three-dimensional views of the data can be seen. A further window might show the distribution of the sampling points on a map and the user is able to select, for example the top 5% of any histogram and immediately see on a map where the sites occur. The visual associations may suggest new hypotheses. See Haslett et al. (1990) for more details.

Cokriging. If there is sufficient data to establish both the variogram of an expensive-to-measure attribute, the variogram of a cheap-to-measure attribute with which it is correlated and the covariogram of their joint variation, then the spatial variation of the expensive attribute can be mapped using fewer samples than otherwise might be the case (McBratney & Webster, 1983b; Stein et al., 1988; Alemi et al., 1988). The method would seem to be of most value when there are good pedogenic or other reasons for expecting the two attributes to have similar spatial patterns (c.f., Leenaers et al., 1989a,b).

Using Models. Many attributes of the soil about which information is required are too expensive or too complex to measure directly and their values are often computed from cheaper, easier-to-measure data using regression models or deterministic models of crop growth, etc. (Burrough, 1989; Dumanski & Onofrei, 1989). When a given model is used with spatial data to compute results for blocks of land, the errors in the model outputs for each block (the prediction errors) accrue from two sources. The first source is the errors in the model parameters; the second is the prediction error of each of the input attributes for each block to the model. Heuvelink et al. (1989) have demonstrated how these errors can be partitioned so that sensible decisions can be made as to whether sampling effort needs to be concentrated on reducing model parameter errors or on reducing the errors of the input data. The first source of error can be improved by using more observations to calibrate or fit the model (particularly if it is a regression model) while the second source of error can be reduced by kriging if the variogram is known and is not pure nugget.

DISCUSSION AND CONCLUSIONS

There is no single approach to sampling. Before sampling, it is important to define the aims clearly and to specify the support and level of spatial resolution needed. Knowledge of the spatial covariance function can help reduce sample numbers considerably, but this function must itself be estimated by sampling. Nested sampling methods are useful for quick estimates of critical scales of spatial dependence; but the variation in the form of an experimental variogram can be large when samples are few. Providing map units of the same class all embrace the same kinds of spatial variation, there seems to be no reason why variograms cannot be computed for sets of map unit occurrences. When attributes are so expensive to measure that it is impracticable to collect enough samples even to determine the variogram, information from limited data can be maximized by using dynamic graphics (c.f., Haslett et al., 1990), cokriging with a cheap-to-measure attribute, or regression models and error propagation.

Conclusions and Suggestions Concerning Optimal Sampling When Spatial Variation is Important

1. If samples are located at intervals greater than a given maximum range r_i of spatial dependence in an area; then the sample set cannot resolve any spatial variation within the area but estimates only the μ and σ^2. This situation is the same as that assumed by the classic model of complete homogeneity. The variance estimated is that accruing from all scales larger than that physically covered by a single sample.

2. Random variation at short ranges can swamp that over longer distances. Consequently in order to investigate the weaker, long-range variation it is essential that samples should only estimate average values of the short-range, more variable components. Short-range variation can be removed by bulking (taking several small samples and computing local averages) or by ensuring that the sample is physically large enough to encompass the small-scale variation. The latter technique may be more suitable when the short-range variation is not normally distributed (e.g., soil pores and cracks). The variance estimated includes only that accrued from scales larger than the samples—measuring larger samples or bulking removes short-range variance.

3. Increasing sampling density may only give limited increases in information when sampling intervals are less than the ranges of dominant spatial scales of variation.

4. Knowing the form of the spatial variation function (the variogram) is critical for optimizing sampling. Knowledge of the spatial variation function allows one to tailor sampling strategies to individual patterns of variation. It also allows decisions to be made about whether expensive-to-measure attributes can be estimated better by (i) direct sampling, or (ii) variation with other cheaper-to-measure attributes.

5. The following experiment could be carried out to see how easy it is to gather useful information about spatial variation from existing soil survey data. For major mapping units it should be possible (i) to compute which attributes are spatially correlated, (ii) to see where supplementary information needs to be gathered, and (iii) to examine how stable the spatial covariance functions (variograms) are for various occurrences of the mapping unit. The results could be evaluated using independent data. A modest effort conducted mainly by the computer could yield immense savings in future

sampling costs, particularly if the results could be made available to soil scientists in the form of sets of guidelines. The tools are now available and should be used.

ACKNOWLEDGMENTS

My thanks are due to R. A. MacMillan (Alberta Research Council), H. ten Berge (Agricultural University Wageningen), A. Bregt (Stiboka, Wageningen) and A. Mateos (Cornell University, Ithaca) for permission to use the data for this chapter and G. Heuvelink for critical comments. Thanks are also due to the referees for suggesting further improvements.

REFERENCES

Alemi, M.H., M.R. Shahriari and D.R. Nielsen. 1988. Kriging and cokriging of soil water properties. Soil Technol. 1:117-132.
Burgess, T.M., and R. Webster. 1980a. Optimal interpolation and isarithmic mapping of soil properties: 1. The semi-variogram and punctual kriging. J. Soil Sci. 31:315-331.
Burgess, T.M., and R. Webster. 1980b. Optimal interpolation and isarithmic mapping of soil properties: 2. Block kriging. J. Soil Sci. 31:333-341.
Burgess, T.M., R. Webster, and A.B. McBratney. 1981. Optimal interpolation and isarithmic mapping of soil properties: 4. Sampling strategy. J. Soil Sci. 32:643-659.
Burrough, P.A. 1983a. Multi-scale sources of spatial variation in soil. I. The application of Fractal concepts to nested levels of soil variation. J. Soil Sci. 34:577-597.
Burrough, P.A. 1983b. Multi-scale sources of spatial variation in soil. II. A non-Brownian Fractal model and its application to soil survey. J. Soil Sci. 34:599-620.
Burrough, P.A. 1986. Principles of geographical information systems for land resources assessment. Clarendon Press, Oxford.
Burrough, P.A., 1989. Matching spatial databases and quantitative models in land resource assessment. Soil Use Manage. 5:3-8.
Cressie, N. 1982. Playing safe with misweighted means. J. Am. Stat. Assoc. 77:754-759.
Cressie, N., and D.M. Hawkins. 1980. Robust estimation of the variogram. J. Math. Geol. 15:115-125.
Davis, J. 1986. Statistics and data analysis in geology, 2nd ed. Wiley, New York.
Dumanski, J., and C. Onofrei. 1989. Crop yield models for agricultural land evaluation. Soil Use Manage. 5:9-15.
Englund, E., and A. Sparks. 1988. GEO-EAS user's guide. Environmental Monitoring Systems Lab., Office of Research and Development, EPA, Las Vegas, NV.
Gilbert, R.O. 1987. Statistical methods for environmental pollution monitoring. Van Nostrand, New York.
Haslett, J., G. Wills, and A. Unwin. 1990. SPIDER—An interactive statistical tool for the analysis of spatially distributed data. Int. J. Geograph. Inform. Syst. 4:285-296.
Heuvelink. G.B.M., P.A. Burrough, and A. Stein. 1989. Propagation of error in spatial modelling with GIS. Int. J. Geograph. Inform. Syst. 3:303-322.
Journel, A.G., and C.J. Huijbregts. 1978. Mining geostatistics. Academic Press, London.
Leenaers, H., J.P. Okx, and P.A. Burrough. 1989a. Co-kriging: An accurate and inexpensive means of mapping floodplain soil pollution by using elevation data. p. 371-382. *In* M. Armstrong (ed.) Geostatistics. Proc. Geostatistics Congr., 3rd. Avignon. 5-9 Sept. 1988. Kluwer, Dordrecht, Netherlands.
Leenaers, H., P.A. Burrough, and J.P. Okx. 1989b. Efficient mapping of heavy metal pollution on floodplains by co-kriging from elevation data. p. 37-50. *In* J. Raper (ed.) Three dimensional applications in geographic information systems. Taylor and Francis Limited, London.
MacMillan, R.A., W.L. Nikiforuk, R.M. Krzanowski, and T.S. Balakrishna. 1987. A soil survey of the Lacombe Experimental Farm, Alberta. Rep. Alberta Res. Council, Edmonton, Canada.

Marsman, B., and J.J. de Gruijter. 1984. Dutch soil survey goes into quality control. p. 127–134. *In* P.A. Burrough and S.W. Bie (ed.) Soil information systems technology. PUDOC, Wageningen.

Mateos, A., P.A. Burrough, and J. Comerma. 1987. Analisis espacial de propiedes de suelo para estudios de modelacion de cultivos en Venezuela. p. 164–178. *In* M. Lyew (ed.) Proc. Primera Conf. Latinoamericana Inform. Geografia, 1st. October 1987. Dep. of Geogr., Univ. of Costa Rica, San José, Costa Rica.

McBratney, A.B., and R. Webster. 1981. The design of optimal sampling schemes for local estimation and mapping of regionalized variables: II. Program and examples. Comput. Geosci. 7:335–365.

McBratney, A.B., and R. Webster. 1983a. How many observations are needed for regional estimation of soil properties? Soil Sci. 135:177–183.

McBratney, A.B., and R. Webster. 1983b. Optimal interpolation and isarithmic mapping of soil properties: V. Co-regionalization and multiple sampling strategy. J. Soil Sci. 34:137–162.

McBratney, A.B., R. Webster, and T.M. Burgess. 1981. The design of optimal sampling schemes for local estimation and mapping of regionalized variables: 1. Theory and method. Comput. Geosci. 7:331–334.

Oliver, M.A., and R. Webster. 1986. Combining nested and linear sampling for determining the scale and form of spatial variation of regionalized variables. Geogr. Anal. 18:227–242.

Riezebos, H. 1989. Application of nested analysis of variance in mapping procedures for land evaluation. Soil Use Manage. 5:25–29.

Rijsuniversiteit Utrecht. 1988. PCGeostat manual. Internal publication. Dep. Physical Geography, the Netherlands.

Ripley, B.D. 1981. Spatial statistics, Wiley, New York.

Rousseeuw, P.J., and A.M. Leroy. 1987. Robust regression and outlier detection. Wiley, New York.

Schiujt, S. 1989. Interpolatie van ruimtelijke gegevens met behulp van het nearest neighbor model. MSc. rep. Rijksuniversiteit Utrecht, Universiteit Twente, The Netherlands.

Stein, A., M. Hoogerwerf, and J. Bouma. 1988. Use of soil map delineations to improve (co)kriging of point data on moisture deficits. Geoderma 43:163–177.

Taylor, C.C., and P.A. Burrough. 1986. Multiscale sources of spatial variation in soil III. Improved methods for fitting the nested model to one-dimensional semivariograms. J. Math. Geol. 18:811–821.

ten Berge, H.F.M., L. Stroosnijder, P.A. Burrough, A.K. Bregt, and M.J. de Heus. 1983. Spatial variability of soil properties influencing the temperature of the soil surface. Agric. Water Manage. 6:213–226.

van Geer, F.C. 1987. Applications of Kalman filtering in the analysis and design of groundwater monitoring networks. Rep. PN 87-05. Inst. Appl. Geosci. TNO. Delft, Netherlands.

Webster, R. 1977. Quantitative and numerical methods in soil classification and survey. Oxford Univ. Press, United Kingdom.

Webster, R. 1985. Quantitative spatial analysis of soil in the field. Adv. Soil Sci. 3:2–70.

Webster, R., and T.M. Burgess. 1984. Sampling and bulking strategies for estimating soil properties in small regions. J. Soil Sci. 35:127–140.

Webster, R., and S. Nortcliff. 1984. Improved estimates of micro nutrients in hectare plots of the Sonning Series. J. Soil Sci. 35:667–672.

Welch, B.L. 1937. The significance of the difference between two means when the population variances are unequal. Biometrika 29:350–362.

8 Presentation of Statistical Data on Map Units to the User

R. B. Brown
University of Florida
Gainesville, Florida

J. H. Huddleston
Oregon State University
Corvallis, Oregon

ABSTRACT

One of the most important facts regarding soil maps, soil map units, and associated text/tables is that they are imprecise. It is well to continue characterization of map units and their variability for the benefit of both the makers and the users of soil inventories. Tabulations of statistics (mean percentages, confidence intervals, confidence levels, etc.) should be presented in soil survey reports if available; but such data must be reported and used cautiously. Simple descriptive statistics do not necessarily lend themselves to probability statements and risk analysis. Users need understandable explanations of reported statistics and their proper use. Users also need improved definitions, explanations, and interpretations of similar and contrasting soils. Close cooperation between users and field soil scientists, educational efforts to assist the public in using soil survey information, and interdisciplinary research to improve understanding of the landscape should take priority over radical changes in format or style of soil survey reports.

The soil landscape is highly variable, and only some parts of that variability are captured in soil survey reports. One of the most important facts regarding soil maps, soil map units, and associated texts/tables, therefore, is that they are imprecise representations of natural landscapes. Users of soil survey reports and related information, be they lay citizens, planners, regulators, agriculturists, or modelers of contaminant behavior in the soil landscape, need to know about both the imprecision of the information available and the variability of the landscape itself (Brown, 1988b; Wilding, 1988).

Copyright © 1991 Soil Science Society of America, 677 S. Segoe Rd., Madison, WI 53711, USA. *Spatial Variabilities of Soils and Landforms.* SSSA Special Publication no. 28.

A poorly characterized map unit is not worth presenting to users, nor for that matter is it worth creating in the first place. It is important to continue characterization of map units and their variability for the benefit of both the makers and the users of soil inventories. Tabulations of statistics (mean percentages, confidence intervals, confidence levels, etc.) should be presented in soil survey reports if available; but such data must be handled with caution. Users need understandable explanations of reported statistics and their proper use.

There are two main types of users, other than field soil scientists themselves, for whom soil surveys are made: the nontechnical and the technical (Brown, 1988b). The degree of complexity and detail required by the two different groups may vary markedly. Farmers and homeowners, for example, often require boiled-down, uncomplicated information on land attributes, either from a soil survey or from an on-site evaluation. Such users might ask, for example, whether or not a mounded septic system is likely to be required to dispose of wastewater adequately. Or they may want to know what sorts of irrigation design and irrigation scheduling are best for the soils on a tract and for the crops to be grown there. Simulation modelers, on the other hand, need highly detailed and specific sorts of information, such as organic C contents of soils, along with the spatial variability of such attributes.

The soil survey report does better at meeting the more general types of need than it does in meeting the more specific needs of the modeler. Even where the modeler can use the *type* of data in a soil survey report, such data frequently lack the kinds of statistics, such as central tendencies and variations around those central tendencies for a variety of soil features, that the modeler needs. Viewed from another perspective, the work of the modeler, especially when it is focused on complex processes involved in such phenomena as transport and transformations of soil contaminants, is often applicable only to tiny areas of land, perhaps no larger than pedons. A model may exist that predicts very well the behavior of a pesticide in an individual pedon, given certain assumptions as to the nature of that pedon and as to management practices and weather. Lacking solid information on the precision of available soil survey information, however, the technical user cannot employ the model to predict precisely the behavior of the pesticide in the larger landscape.

For the sake of all users, and just as importantly for pedologists' improved understanding of the soil landscape, we should move ahead with efforts to characterize map units statistically to the fullest extent possible. What is reasonable to expect in this regard from the field soil scientist, and how best to package the resulting statistics for the user, are the subjects of the following sections.

DESCRIBING SOIL VARIABILITY FOR THE USER

Knowing the properties of map units has always been one of the highest priorities of the field soil scientist. Increasingly, it is also becoming a high

STATISTICAL DATA ON MAP UNITS

priority for those soil scientists and others who do *not* map, classify, and interpret soils in the field.

It is possible to know virtually everything about the variability of a soil map unit, but to gain that level of knowledge would require the making of maps that approach 1:1 in scale and the expenditure of unreasonable amounts of time and money. Nevertheless, progress can be made toward improved understanding of map unit composition and variability. Routine collection of statistical data by transect or other types of sampling is one means toward this end. Two of the benefits of having transect data routinely collected and compiled into statistics are:

1. Quality control is improved as a survey progresses. Inaccurate or otherwise poor mapping and/or correlation can be discovered and corrected early in the survey. Map units that need to be redefined, split, combined, or remapped can be dealt with accordingly. Appropriate names can be assigned to map units to reflect certain types of variability. Decisions can be made as to how best to handle extreme sorts of variability (as with complexes and undifferentiated groups) in naming and interpreting the map units.
2. The final product (the soil survey report and any ancillary information) is improved. The report itself is better both for the reasons given in Item 1 above and because the transect data help the user to understand the imprecision of the report and, even more importantly, the variability and nature of the soil landscape itself.

Quantifying Variability

National Cooperative Soil Survey (NCSS) reports are rigidly standardized as to format and content. Our intent in this chapter is not to attack that rigidity. In fact, we are inclined to support its maintenance, with minor adjustments to improve truthfulness, readability, and utility of the reports. The use of soil survey reports and related information will be advanced to a greater degree by effective educational and interpretive programs, and by relevant, interdisciplinary research, than by radical alterations of format and content of the reports.

Within the current constraints of format and content, there have been some improvements in the usefulness of soil survey reports. We now turn to some examples of recent innovations in such reports, with particular emphasis on the reporting of soil variability. Five reports published in the last 10 yr are taken as examples.

For the *Soil Survey of Huron County, Michigan* (Linsemier, 1980), transect data were collected and tabulated in the report (Table 8–1A). This tabulation certainly helps the thoughtful user to grasp the variability of the map units. The data are not, however, employed elsewhere in the report, at least not in a way that is obvious to the user. Narrative map unit descriptions from the same report (Table 8–1B) make no reference to the transect data, and even omit mention of the rough percentages of inclusions that are

Table 8-1A. Average composition of selected map units as tabulated in the *Soil Survey of Huron County, Michigan* (Linsemier, 1980).

Symbol and soil name	Observations	Named and similar		Somewhat contrasting		Strongly contrasting	
	no.	series	%	series	%	series	%
3A Shebeon loam, 0 to 2% slopes	280	Shebeon Similar†	34 25	Kilmanagh Aubarque Sanilac Other	16 6 6 7	Avoca Other	3 3
4B Grindstone loam, 0 to 4% slopes	93	Grindstone	73	Shebeon Gagetown Other	10 7 4	Avoca Other	3 3
51B Guelph-Londo loams, 2 to 6% slopes	115	Guelph Londo	48 40	Parkhill Other	3 5	Other	4
56A Riverdale-Pipestone, 0 to 2% slopes	69	Riverdale Wasepi Pipestone	43 15 25	Rapson	11	Londo Tobico	3 3

† Similar to Shebeon soil, but no argillic horizon was discernible by field observations.

STATISTICAL DATA ON MAP UNITS

Table 8-1B. Selected statements of map unit compositions from detailed map unit descriptions in the *Soil Survey of Huron County, Michigan* (Linsemier, 1980).

3A, Shebeon loam, 0-2% slopes

... Included with this soil in mapping are small areas of poorly drained Kilmanagh soils and somewhat poorly drained Avoca soils. The Kilmanagh soils are in shallow depressions and drainageways. The Avoca soils are coarser textured than the Shebeon soil and are throughout the unit. Also included are small areas of calcareous Aubarque and Sanilac soil. A few small areas in which bedrock is at a depth of less than 60 inches are throughout the map unit....

4B, Grindstone loam, 0-4% slopes

... Included with this soil in mapping are small areas of somewhat poorly drained Shebeon and Avoca soils. They are in depressions and narrow drainageways. Also included are moderately well drained Gagetown soils which are coarser textured and have higher percentage of silt and very fine sand than the Grindstone soils. They are interspersed throughout the unit. A few small areas in which bedrock is at a depth of less than 60 inches are throughout the map unit....

51B, Guelph-Londo loams, 2-6% slopes

... Included with these soils in mapping are small areas of poorly drained and very poorly drained Parkhill soils. They are in the most prominent depressions and drainageways....

56A, Riverdale-Pipestone complex, 0-2% slopes

... Included with these soils in mapping are small areas of Londo soils which are finer textured than the Riverdale and Pipestone soils and small areas of Rapson soils that are finer textured in the substratum. They occur throughout the unit. Also included are poorly drained and very poorly drained Tobico soils. They are in depressions and drainageways....

so common in the map unit descriptions of NCSS soil survey reports. Linkage between the tabulated and narrative data, definitions of the terms used in the tabulations (somewhat contrasting, etc.), and discussions of the data's implications would have increased the value of the information for the user.

The NCSS program definitely could benefit from improved definitions of similar and contrasting soils. The *National Soils Handbook* (Soil Survey Staff, 1983) advises mappers to consider interpretive as well as taxonomic differences in deciding whether an inclusion is similar or dissimilar to the named soil(s). We may be comfortable with our own reasoning and decisions; but have we conveyed adequately to users our own definitions of inclusions? If inclusions mean one thing to us and another thing to our clientele, then our clientele are not as well served as we may think.

The *Soil Survey of Jasper and Newton Counties, Texas* (Neitsch, 1982), like the Huron County report, contains transect data; but here the data are not tabulated. Rather, they are folded into the map unit descriptions of complexes, associations, and undifferentiated groups (Table 8-2). Descriptions of consociations (a minority of the map units in these two heavily forested counties) do not include transect data in this report; the reported percentages of inclusions in consociations are based instead on the judgment of the mapper. The descriptions of the complexes, associations, and undifferentiated groups (Table 8-2), on the other hand, are based on transect data and are quite revealing and helpful as to the actual variability of the map units.

Table 8-2. Selected statements of map unit compositions from detailed map unit descriptions in the *Soil Survey of Jasper and Newton Counties, Texas* (Neitsch, 1982).

AtA, Attoyac fine sandy loam, 0–3% slopes

... Included with this soil in mapping are small areas of Bernaldo, Besner, and Mollville soils. Bernaldo soils are on landscape positions similar to those of Attoyac soils. Besner soils are on mounds. Mollville soils are in small depressions, mostly of less than two acres. A soil similar to the Attoyac soil but having a fine sandy loam subsoil is on low ridges. These included soils makeup about 20 percent of any mapped area. ...

BeB, Besner-Mollville complex, gently undulating

... This complex is 50 to 60 percent Besner soils, 30 to 35 percent Mollville soils, and 10 to 15 percent other soils. These percentages are determined by taking samples from random transects made across mapped areas. ...

DUB, Doucette-Boykin association, undulating

... This association is 30 to 35 percent Doucette soils, 15 to 40 percent Boykin soils, and 20 to 30 percent other soils. These percentages were determined by taking samples from random transects made across mapped areas. ...

Mn, Mantachie and Bleakwood soils, frequently flooded

... This undifferentiated group is 45 to 65 percent Mantachie soils, 25 to 40 percent Bleakwood soils, and 5 to 10 percent other soils. These percentages were determined by taking samples from random transects made across mapped areas. ...

Elsewhere in this soil survey report there are clear definitions of each of these types of map unit. There also is a section addressing confidence limits of soil survey information. Including such a section was and is both innovative and beneficial; but there is some language employed in the description of confidence limits that may be misleading. More discussion of this problem appears in a subsequent section of this chapter entitled "Confidence Levels."

In *Soil Survey of Conecuh County, Alabama* (Fox, 1989), narrative descriptions of both single- and multinamed map units contain clear indications of which mentioned inclusions are contrasting and what their estimated percentages are (Table 8-3). This style of reporting inclusions is easy to do and understand. Too many other reports (not referenced here) leave the user to infer or guess whether the reported percentages are meant to include contrasting inclusions or contrasting plus similar inclusions. Another good feature of map unit descriptions in the Conecuh County report, and many others, is the use of a semitabular format, rather than sentences, to report those soil attributes that lend themselves to succinct assemblage, for example:

... Permeability: moderate
Available water capacity: moderate
Reaction: very strongly acid to medium acid
Organic matter content: low
Natural fertility: low
Depth to bedrock: more than 60 inches
Root zone: to a depth of more than 60 inches
High water table: none within a depth of 72 inches
Flooding: none . . . (Fox, 1989).

Table 8-3. Selected statements of map unit compositions from detailed map unit descriptions in the *Soil Survey of Conecuh County, Alabama* (Fox, 1989).

BgA, Bigbee sand, 0-1% slopes, rarely flooded

... Included with this soil in mapping are a few areas of Bonneau, Cahaba, Chrysler, and Izagora soils. The included soils make up about 25 percent of the map unit, but individual areas are generally less than 10 acres. These included soils are contrasting soils, and their use and management differ from those of Bigbee soil. ...

CbA, Cahaba-Bigbee complex, 0-2% slopes, rarely flooded

... Included in mapping are areas of Bonneau, Chrysler, and Yonges soils. Also included are soils that are similar to Cahaba soil, but that have a yellowish brown subsoil. The included soils make up about 15 percent of the map unit, but individual areas are generally less than 10 acres. Chrysler and Yonges soils are contrasting soils, and their use and management differ from those of Bigbee and Cahaba soils. The contrasting soils make up about 10 percent of the map unit. ...

PoB, Poarch sandy loam, 0-5% slopes

... Included with this soil in mapping are a few small areas of Fuquay, Malbis, and Troup soils. Also included are soils that are similar to Poarch soil except they have either a loamy sand surface layer or are somewhat poorly drained. The included soils make up about 15 percent of the map unit, but individual areas are generally less than 5 acres. Fuquay and Troup soils are contrasting soils, and their use and management differ from those of Poarch soil. The contrasting soils make up about 5 percent of the map unit. ...

This style is efficient and easy to read. The Conecuh County report is an example of a report that deviates little from the standard format of NCSS; but the little changes that *are* made prove very helpful to the user.

The *Soil Survey of Decatur County, Kansas* (Hamilton et al., 1989) is an example of a report that lacks any indication that transect data were employed in map unit design or in estimates of inclusions. We are left to infer that mappers' judgment and/or transect data were the sources of information on inclusion percentages (Table 8-4). The percentages reported in the map unit descriptions do range widely, however, and the astute user can gain good understanding of which are the more uniform and which are the more variable segments of the landscape of this county. Percentages of inclusions that range from "... less than 1 percent ..." in one map unit, to "... about 25 percent ..." in another, suggest an intent on the part of the report writers honestly to convey the nature of the landscape as they understand it.

This wide range of stated purities stands in contrast with ranges in many other reports, even some of the more recent, which tend to stay in or near the 5 to 15% range, reflecting perhaps some disinclination to report the truth. Such reticence may in the end diminish both the credibility and utility of such reports.

In the *Soil Survey of Hillsborough County, Florida* (Doolittle et al., 1989), we come again to a tabular presentation of transect data. Unlike the Huron County, MI report (Linsemier, 1980), however, the data here are reported with confidence limits and confidence levels (Table 8-5A). These variability data are also discussed elsewhere in the Hillsborough report, in

Table 8-4. Selected statements of map unit compositions from detailed map unit descriptions in the *Soil Survey of Decatur County, Kansas* (Hamilton et al., 1989).

Bd, Bridgeport silt loam, 0-2% slopes
... (No statement regarding inclusions) ...
Ha, Holdrege silt loam, 0-1% slopes
... Included with this soil in mapping are small areas of moderately well drained Pleasant soils in shallow depressions. These soils make up less than 1 percent of the map unit. ...
Mu, Munjor sandy loam, occasionally flooded
... Included with this soil in mapping are small areas of the silty McCook soils. These soils are in positions on the landscape similar to those of the Munjor soil. They make up about 15 percent of the map unit. ...
Ph, Penden-Canlon loams, 6-30% slopes
... Individual areas ... are about 45 percent Penden soil and 30 percent Canlon soil. The two soils occur as areas so intricately mixed or so small that mapping them separately is impractical. ... Included with these soils in mapping are small areas of the deep, silty McCook and Coly soils and small areas of rock outcrop. The rock outcrop is caliche. It is on the steeper side slopes. Included areas make up about 25 percent of the map unit. ...

sections on use of ground-penetrating radar (GPR) and on confidence limits of soil survey information, which offer generic descriptions of how the data were collected and what they mean. The data also appear in the language and percentages of inclusions given in the narrative map unit descriptions (Table 8-5B). Transect data must have been very helpful in assigning names to map units for this soil survey report, which is an update of a report published in the 1950s, and in deciding which map units should be consociations vs. associations, etc. The data are certainly helpful to the user in understanding the imprecision of the report and in understanding the landscape. In its detailed reporting of rigorously determined map unit compositions, this report takes another in the series of major steps forward that began with the Huron County report and continued with the report for Jasper and Newton Counties.

Confidence Levels

As with the soil survey report for Jasper and Newton Counties, TX (Neitsch, 1982), the language employed in the Hillsborough County, FL report to describe the meaning of the transect-derived statistics does not seem to be entirely correct. Each of these two reports has a section entitled "Confidence Limits of Soil Survey Information." In the Jasper and Newton County report, that section reads, in part:

> The statements about soil behavior in this survey can be thought of in terms of probability: they are predictions of soil behavior. ... Confidence limits of soil surveys are statistical expressions of the probability that the composition of a map unit or a property of the soil will vary within prescribed limits. Confidence limits can be assigned numerical values based on a random sample. ... Soil scientists made enough transects and took enough samples to

STATISTICAL DATA ON MAP UNITS

Table 8-5A. Average composition of selected map units as tabulated in the *Soil Survey of Hillsborough County, Florida* (Doolittle et al., 1989).

Symbol and soil name	Transects	Soils		Confidence interval	Confidence level	Dissimilar soils	
	no.		%	%	%		%
2 Adamsville fine sand	11	Adamsville	86	82–99	95	Lochloosa	6
		Similar soils	7			Pomello	1
17 Floridana fine sand	4	Floridana	54	80–98	60	Samsula	6
		Similar soils	36			Wabasso	4
46 St. Johns fine sand	16	St. Johns	67	76–99	80	Basinger	6
		Similar soils	20			Other	7
53 Tavares-Millhopper fine sands, 0–5% slopes	14	Tavares	50	87–99	95	Myakka	5
		Similar soils	13			Smyrna	4
		Millhopper	15			Candler	2
		Similar soils	11				

Table 8-5B. Selected statements of map unit compositions from detailed map unit descriptions in the *Soil Survey of Hillsborough County, Florida* (Doolittle et al., 1989).

2, Adamsville fine sand

... In 95 percent of the areas mapped as Adamsville fine sand, the Adamsville soil and similar soils make up 82 to 99 percent of these mapped areas. Dissimilar soils make up 1 to 18 percent of the mapped areas. ... Similar soils included in mapping are very dark grayish brown or dark grayish brown fine sand in the lower part of the underlying material. Other similar soils, in some of the higher parts of the landscape, are moderately well drained. ... Dissimilar soils included in mapping are Lochloosa and Pomello soils in small areas. ...

17, Floridana fine sand

... In 60 percent of the areas mapped as Floridana fine sand, the Floridana soil and similar soils make up 80 to 98 percent of the mapped areas. Dissimilar soils make up 2 to 20 percent of the mapped areas. ... Similar soils included in mapping, in some areas, have a surface layer that is less than 10 inches thick. Other similar soils have a subsoil within 20 inches of the surface; and in some places, the included similar soils have a subsoil at depth of more than 40 inches. ... Dissimilar soils included in mapping are Samsula and Wabasso soils in small areas. Samsula soils are organic, and Wabasso soils are poorly drained. ...

46, St. Johns fine sand

... In 80 percent of the areas mapped as St. Johns fine sand, the St. Johns soil and similar soils make up 76 to 99 percent of the mapped areas. Dissimilar soils make up 1 to 24 percent of the mapped areas. ... Similar soils included in mapping, in some areas, have a surface layer that is less than 10 inches thick. Other similar soils, in some places, do not have a subsurface layer; and in some places, these included soils have a subsoil that is brown or dark brown. ... Dissimilar soils included in mapping are Basinger soils in small areas. Basinger soils are very poorly drained. Also included are unnamed soils that have a surface layer that is 10 to 24 inches thick and have a loamy layer at a depth of more than 40 inches. ...

53, Tavares-Millhoper fine sands, 0-5% slopes

... In 95 percent of the areas of this map unit, Tavares-Millhopper fine sands, 0 to 5 percent slopes, and similar soils make up 87 to 99 percent of the mapped area, and dissimilar soils make up 1 to 13 percent of the mapped areas. Generally, the mapped areas consist of about 63 percent Tavares soil and similar soils and 26 percent Millhopper soil and similar soils. ... Similar soils [to Tavares] included in mapping, in some areas, have a brown or dark brown layer in the lower part of the underlying material. Other similar soils, in some of the lower parts of the landscape, are somewhat poorly drained. ... Similar soils [to Millhopper] included in mapping, in some areas, have a dark surface layer more than 10 inches thick. ... Dissimilar soils which are included in this map unit are Candler, Myakka, and Smyrna soils in small areas. Candler soils are excessively drained. Myakka and Smyrna soils are poorly drained. ...

characterize the delineated associations and complexes at an 80 percent confidence level. This means, for example, that in 80 percent of the area mapped as Newco-Urland association, hilly, the percentage of the soils will be within the range given in the map unit description. In as many as 20 percent of the mapped areas of this association, the percentage of any of the soils can be either higher or lower than the given range. (Neitsch, 1982)[1]

---

[1] See APPENDIX for family designations of all soil series mentioned in this chapter.

The corresponding section in the Hillsborough County report contains similar language. In fact, the narrative confidence statements from Neitsch (1982) were adopted, with slight modifications, for use in Florida (Schellentrager et al., 1988). The Hillsborough report reads, in part:

> Confidence limits are statistical expressions of the probability that the composition of a map unit or a property of the soil will vary within prescribed limits. Confidence limits can be assigned numerical values based on a random sample. . . . Specific confidence limits for the composition of map units in Hillsborough County were determined by random transects made with the GPR across mapped areas. . . . Soil scientists made enough transects and took enough samples to characterize each map unit at a specific confidence level. For example, map unit 29 was characterized at a 95 percent confidence level based on the transect data. The resulting composition would read: In 95 percent of the areas mapped as Myakka fine sand, Myakka soil and similar soils will comprise 84 to 93 percent of the delineation. In the other [5] percent of the areas of this map unit, the percentage of Myakka soil and similar soils may be higher than 93 percent or lower than 84 percent (Doolittle et al., 1989).

We accept the data that these two reports provide on map units. What we question is the way these data are presented in the above-quoted and other narrative sections of the reports. The confidence level given for each map unit should be stated as the degree of confidence that can be assigned to the *sampling* that was done in that map unit (Snedecor & Cochran, 1980; Devore & Peck, 1986; Ostle & Malone, 1988). A set of samples from transects conducted on a map unit has a specified confidence level, meaning that the confidence interval resulting from that sampling event has the specified probability of including the real population value. In the case of the Myakka map unit used as an example above, the 95% confidence level really means: the probability was 0.95 that the interval estimated by sampling along random transects actually would include the true population percentage of Myakka plus similar soils. If the same sort of sampling had been used in many different and separate sampling experiments on the land mapped as Myakka in Hillsborough County, approximately 95% of the resulting confidence intervals would have included the real population percentage of Myakka plus similar soils. About 5% of such sampling efforts, however, would have yielded intervals that did *not* include the true percentage. We do not know whether the one sampling effort conducted on Myakka in this county is one of the 95% of such efforts that would include the true percentage or one of the 5% that would *not*!

Given the subtle but important distinctions among the different ways of expressing confidence levels, and given the widely ranging backgrounds and levels of receptiveness of users, it probably would be best to list the confidence levels only once—perhaps in a report's tabulation of transect data. A footnote to the table could describe what the confidence levels mean. But no mention needs to be made in the narrative map unit descriptions. A map unit description might thus read as follows:

... Myakka soil and similar soils make up approximately 84 to 93 percent of the areas mapped as Myakka fine sand. Dissimilar (contrasting) soils make up the remaining 7 to 16 percent. ...

From here the map unit narrative should continue directly into a description of where the dissimilar soils are likely to be found within the areas mapped as Myakka fine sand, as well as how they differ from the Myakka soil. Are the inclusions small or large? Do they have a characteristic shape? Do they occur associated with readily observable landscape features? Do inclusions occur in a similar manner in every delineation? Are they dominantly near the edges of delineations? Are any inclusions clustered in one corner of the county? Does the percentage shift from east to west? Does it shift as one gets farther from a certain body of water or other feature? In what important way(s) are the inclusions contrasting? Answers to questions like these provide the kinds of information that help the user to understand soil landscapes. Such statements are just as vital, perhaps more so, in narratives for multitaxa map units. It is important for mappers themselves to distinguish properly among consociations, associations, complexes, and undifferentiated groups. It is just as important for *users* to know what the mappers did and why they did it.

There is room for debate as to (i) what confidence level should be selected (80%, 90%, 95%, etc.), and (ii) whether the confidence level should be held constant (perhaps at 90%) among all map units in a survey area (R.W. Arnold, 1989, personal communication). It is argued by some that allowing the confidence level to drift among map units gives the impression that the mappers have dropped the confidence level for some highly variable map units in order to narrow the confidence intervals and make them more presentable. On the other hand, time or other constraints may prevent sufficient sampling to achieve meaningful, useful confidence intervals at a rigidly specified confidence level for every map unit in a survey area. The team in the field may be best qualified to decide which statistic, the confidence interval or the confidence level, should best be allowed to slide.

It has also been pointed out (R.W. Arnold, 1989, personal communication) that many users are really only interested in the *lower* end of the confidence interval, since this is the more pessimistic (i.e., the least pure) of the two extremes defining the confidence interval. This lower end of the estimate therefore best identifies the "risk" of using the map as published without further on-site evaluation. Some report writers may want simply to report this lower level by itself, ignoring the sample mean and the high end of the confidence interval. Doing so would have the added advantage of allowing the confidence level to be increased, as from 80 to 90%, or from 90 to 95%; because only a one-sided confidence interval would now be employed and reported (Wonnacott & Wonnacott, 1985).

Collection of Data

It is certainly wise, for a variety of reasons, to undertake statistical sampling of map units and tabulation of the results for the benefit of mappers and users. There may be some pitfalls, however, in expecting the soil survey party to do the sampling and data workup. First, the party members may have acreage goals as a nagging concern. Second, they may be protective, and rightfully so, of their ability to read the soil landscape and convey their findings to users employing proven, time-honored principles of pedology. Third, mappers are naturally inclined to put the best light on their own work. They are apt to be biased, not only in selecting where to make their observations for a variability study, but also in deciding whether certain of the findings might better be left unreported. Under such constraints, transect studies conducted by the mappers themselves may not always receive the rigor that such sampling demands.

For a less-biased sample, data might best be gathered, concurrently with a progressive soil survey, by a team of outsiders, perhaps a "floating" soil survey transecting party that has training in and primary responsibility for conducting transect and/or other types of sampling and statistical analysis. Granted, such a group might itself have biases. Such workers will want to be supportive of the profession of field soil scientists. And even if they were perfectly unbiased in attitude, they will of necessity be biased whenever they are unable to sample in certain inaccessible areas (e.g., under freeways, under tree trunks, in yards bearing "Beware of Dog" signs, etc.). But they are less likely to have a self-serving sort of bias than the individuals actually drawing the lines and assigning symbols.

It is said among field soil scientists that one is not ready to begin mapping an area until one has finished mapping the area. The point is that the process of soil map legend building is not really complete until all of the area to be mapped has been studied fully. When modern soil surveys get updated to meet changing needs of users, as in rapidly urbanizing areas, heavy responsibility for transecting should be placed on the update teams as they evaluate both the mapping that has been done and the legend that has been used to conduct that mapping. They should have the latitude to alter the old legend accordingly and to describe in an unbiased fashion the nature of the map units, in a language that best reflects the nature of the soil landscape.

Beyond tabular and/or narrative reporting of confidence intervals and confidence levels for inclusions within map units, we do not believe that soil survey report writers should go very far toward risk analysis or probability statements. The main reason is that we simply don't know at present what the user should or would be able to do with such information. If, for example, we tell users that there is approximately a 70 to 90% chance of finding the named soil in a map unit, or that the odds are 65 to 75% that the soil permeability will be adequate for a septic system, what is the user going to do with that information? Unless we can provide specific guidance on how to conduct or take advantage of risk analysis or probability statements, and unless there is good statistical justification for making such analyses or statements in specific contexts, we would be opposed to putting very much of that kind of information in soil survey reports.

It is argued increasingly by soil interpreters that we err in interpreting only pedons or, perhaps more correctly, in interpreting phases of dominant taxonomic units within map units, when we develop interpretive tables for soil survey reports. This is at least partially true. It remains to be shown, however, just how variability itself [intricacies of spatial arrangement, degree(s) of interpretive contrast, sizes of the contrasting landscape segments, etc.] can best be incorporated into the interpretive decision tree. And it continues to be useful and meaningful to the astute soil survey report user to see tabulated the interpretations of contrasting soils in multitaxa map units. For now, it would seem to be best to let the "on-scene commander" (i.e., the soil survey party leader, in consultation with party members and clientele, with due consideration given to the nature of the local landscape) exercise good judgment in bringing soil variability into the process of interpreting soil map units.

The same could be said regarding the description and interpretation of some other soil attributes, including temporal variations. The manner in which water table dynamics are reported, for example, ranges widely around the nation. Many reports contain narratives that simply restate information on the wet-season water table's "depth, kind, months" from the Soil and Water Features table of the same reports. Others contain additional information that reflects particular emphasis on water tables by the makers and local users of the report. The Hillsborough County, FL report (Doolittle et al., 1989), for example, offers insights as to depth and duration of wet-season water table for some of its more important map units. The water table of St. Johns fine sand is reported as follows:

> In most years, a seasonal high water table fluctuates from the soil surface to a depth of 15 inches for 2 to 6 months and recedes to a depth of 15 to 30 inches during prolonged dry periods.

Tavares soil, on the other hand,

> . . . has a seasonal high water table at a depth of 40 to 80 inches for more than 6 months, and it recedes to a depth of more than 80 inches during prolonged dry periods.

Millhopper [which occurs in association with Tavares (Table 5A)], is reported to have

> . . . a seasonal high water table at a depth of 40 to 60 inches for 1 to 4 months, and it recedes to a depth of 60 to 72 inches for 2 to 4 months.

Again, it seems best to let soil survey parties decide themselves what the local needs are and how confident they feel about the information they possess.

To further emphasize to the user the nature of soil variability and the imprecision inherent in a soil survey report, it might help to add a section to the report that includes a step-down sequence of soil maps, each encompassing the same small tract of land, and to show visually the importance of scale in mapping as well as in display of soil information. The maps could progress in successively larger scales from the General Soil Map to the photo-based map sheets of the report to an even larger scale, perhaps 1:1200, that illustrates the kind of detail that can be shown with intensive, site-specific mapping. For maximum credibility and educational value, the step-down sequence of soil maps should show an area well-known to most users (perhaps a park or natural area), and the mapping shown at all the scales must be "real mapping," performed at whatever intensity necessary to yield the information desired. Such a display would help users grasp the fact that there is necessary generalization in virtually any soil map of scale smaller than 1:1. Something similar to this suggested step-down sequence was done for the *Soil Survey of Brazos County, Texas* (Mowery et al., 1958), in which a small area of Lufkin-Edge complex, 3 to 8% slopes, is shown mapped at 1:7920 as part of that particular map unit description. What we envision is a more prominent treatment of the subject, with the step-down maps shown all on one page and with accompanying wording to reinforce the idea that cartographic generalization is necessary for *all* map units in a soil survey.

Ancillary "Fact Sheets" or "Advisory Letters"

Our feeling is that the format of NCSS reports should not be altered greatly at this stage in the current generation of reports. It may be appropriate, however, to supplement the report with fact sheets or other kinds of information in a form that is readily available to the user and whose connection to the published soil survey report is obvious. What we have in mind is a continuing series of special-purpose "letters," "advisory letters," "fact sheets," or the like from which the user may choose when picking up a soil survey report in an agency office, or which could be mailed or handed out to audiences or individual users depending on the subject of concern. These supplements should contain short but informative discussions of particular topics and questions that are recurring and of particular interest to local citizens. Topics might include:

- Accuracy and precision in soil survey
- The wet season water table
- Soil survey digitizing and geographic information systems
- Effects of human activities on the water table
- Soils and stormwater retention facilities
- Laboratory data in soil survey reports—use and misuse
- Why soil names sometimes change across county lines.

Supplements should be state-specific and linked to published reports for the state. The supplement masthead could reflect the format of soil survey report covers in the state, and also call attention to the interagency cooperation inherent in soil survey. An appropriate masthead might look like this:

Interagency Soil Survey Applications Work Group– (land grant university) and USDA-SCS in cooperation with (agencies, professional organizations, others)	Soil Survey of (state)

Soil Survey Advisory Letter no. _____

(Title)

by

(Authors)

.... (Text) ....
....
....
....(etc.)

What we envision here is a set of supplemental fact sheets that highlight the cooperative nature of soil survey and provide locally inspired, timely, quick turnaround, short, readable fact sheets that augment the soil survey report itself; and introduce the profession of soil scientists, public and private, that is available to help where existing information is inadequate to meet particular needs. A likely agency to undertake publication of such a series of fact sheets for a state is the cooperative extension service, in close cooperation with the entire profession of field soil scientists throughout the state.

Serving the More Technical User

Most of the above discussion involves meeting the needs of the nontechnical user who is not highly quantitative and/or who is not likely to have skills in the use of computerized soil data bases and geographic information systems. There certainly are needs in these latter, more technical areas, however. Such needs must be met with basic and applied research concerning the nature of the landscape itself. Much that is exciting is ongoing in these areas. For example, Horvath et al. (1987) report on the combination of a database, remotely sensed landscape data, terrain data, and soils information to aid soil survey makers and users. Robert and Anderson (1987) and several other groups have developed soil survey information systems that are both user-friendly and suitable for work on desktop computers.

In somewhat more basic investigations, some interdisciplinary studies have also yielded results with implications for the field soil scientist. Hendrickx et al. (1988), for example, found that in order for soil surveys to be useful for water and solute transport studies, they should routinely include soil physical measurements *and* sensitivity analyses to determine the accuracy with which horizon thickness must be measured in the course of the survey. Di et al. (1989) showed that there are occasions when geostatistics can improve efficiency and reduce the numbers of samples necessary to carry out soil surveys. Their findings are unlikely to impact progressive second-order soil survey, but special purpose, high-intensity mapping would do well to incorporate their findings.

Stein et al. (1988), in an innovative and promising attempt to capitalize upon the knowledge contained in soil survey reports, employed kriging and cokriging to predict 30-yr average moisture deficits, with and without stratification by soil map delineations. Results showed the value of stratifying soil map information as to soil taxa or as to groundwater class. Their results have what they call "interesting consequences" for soil survey. When a specific degree of accuracy of prediction is required in an experiment involving sampling, it would be feasible to decrease the number of observations in strata showing relatively low variability and to increase the number of observations in strata showing relatively high variability. As Stein et al. (1988) observe,

> In many countries systematic soil surveys are completed. One possible future activity would be to determine the internal variability of existing major land units so as to allow statistically founded quantitative predictions of relevant land qualities rather than qualitative estimates based on the properties of 'representative' profiles.

These studies point up the value of and the need for improved quantitative understanding of the variations that occur within map units. The benefits will accrue to technical and nontechnical users alike. As Bouma (1986) has pointed out, researchers with their analytical, quantitative leanings often come up with findings that are applicable only to small volumes of soil, perhaps the size of a pedon or, worse, the size of a laboratory column. The possessor of a more qualitative sort of insight, such as a farmer, may not have actual numbers, but may have experiences and understanding applicable to an entire region. The field soil scientist is or ought to be conversant with both types of individuals, providing avenues of communication between them, so that each will benefit from the other's experiences, insights, and skills.

All these studies, and others like them, speak to the relationships among (i) what we now know about the landscape, (ii) what more we need to know, and (iii) how to get from "i" to "ii" efficiently. Implied between the lines is a need for interdisciplinary, cooperative, innovative efforts to bring our various forms of knowledge and skills (especially the two extremes of field-oriented and laboratory/mathematical/computer-oriented knowledge and skills) together to address real-world problems.

THE PROFESSION OF FIELD SOIL SCIENTISTS

Soil surveyors operate in a highly complex environment, even apart from the occasionally imponderable landscape itself. The field soil scientist needs increasingly to deal not just with private citizens and planners, but with lawyers who may question the reliability of the product, with statisticians and modelers who need numbers and who may or may not have a true understanding of landscapes, and with other scientists who want to undertake sophisticated, interdisciplinary, scientific investigation and thereby address pressing practical problems. In this environment, the soil scientists who map soils and package their findings to meet users' needs must be at once *scientists*, *professionals*, and *good communicators* (Brown, 1988a,b). They cannot be any of these things in a vacuum.

Members of the Soil Science Society of America and corresponding soil science societies worldwide, most especially those of us affiliated with Division S-5 and its counterparts elsewhere, rightfully claim the center ground of soil science. To claim that territory does not mean to erect barricades around our subdiscipline. It means, most assuredly, the opposite.

To begin with, we have a responsibility to help prepare soil scientists to work in the field, where soils occur and where the clientele for soil survey are found. It is also our responsibility to continue to nurture field soil scientists as professional practitioners of our science throughout their careers (Miller & Brown, 1987; Brown, 1988b; Brown & Miller, 1989).

Finally, we must bring our own and our field-oriented colleagues' knowledge, experience, and skills to modelers and other scientists in useful form. The standard soil survey report will not always (perhaps never) be a perfect vehicle for this purpose, except insofar as it (i) contains an important and central part of our knowledge *and* (ii) serves as the catalyst for our further learning. Interdisciplinary work is not necessarily easy to initiate and carry through to completion (Miller, 1983). It is the only way, however, that many current problems will be addressed efficiently and effectively.

It is no small feat to get the attention of the pragmatic field soil scientist, the potential soil survey user, the experimental scientist, or the modeler. But these parties need to be in communication with one another for soil survey to reach its maximum potential as a means to understanding the soil landscape. If pedologists don't provide the avenues and the environment for such communication to take place, so that all parties to land-use decisions will be functioning with the facts, and so that gaps in knowledge will be filled, it is doubtful that such communication and cooperation will occur at all.

APPENDIX

Family designations, as listed in U.S. Department of Agriculture-Soil Conservation Service (1990), of the soil series named in this chapter.

Series	Family
Adamsville	Hyperthermic, uncoated Aquic Quartzipsamments
Attoyac	Fine-loamy, siliceous, thermic Typic Paleudalfs
Aubarque	Coarse-loamy, mixed (calcareous), mesic Aeric Haplaquepts
Avoca	Sandy over loamy, mixed, mesic Entic Haplaquods
Basinger	Siliceous, hyperthermic Spodic Psammaquents
Bernaldo	Fine-loamy, siliceous, thermic Glossic Paleudalfs
Besner	Coarse-loamy, siliceous, thermic Glossic Paleudalfs
Bigbee	Thermic, coated Typic Quartzipsamments
Bleakwood	Fine-loamy, siliceous, acid, thermic Typic Fluvaquents
Bonneau	Loamy, siliceous, thermic Arenic Paleudults
Boykin	Loamy, siliceous, thermic Arenic Paleudults
Bridgeport	Fine-silty, mixed, mesic Fluventic Haplustolls
Cahaba	Fine-loamy, siliceous, thermic Typic Hapludults
Candler	Hyperthermic, uncoated Typic Quartzipsamments
Canlon	Loamy, mixed (calcareous), mesic Lithic Ustorthents
Chrysler	Clayey, mixed, thermic Aquic Paleudults
Coly	Fine-silty, mixed (calcareous), mesic Typic Ustorthents
Doucette	Loamy, siliceous, thermic Arenic Plinthic Paleudults
Edge	Fine, mixed, thermic Udic Paleustalfs
Floridana	Loamy, siliceous, hyperthermic Arenic Argiaquolls
Fuquay	Loamy, siliceous, thermic Arenic Plinthic Kandiudults
Gagetown	Coarse-silty, mixed, mesic Typic Hapludolls
Grindstone	Fine-loamy, mixed, mesic Glossaquic Hapludalfs
Guelph	Fine-loamy, mixed, mesic Glossoboric Hapludalfs
Holdrege	Fine-silty, mixed, mesic Typic Argiustolls
Izagora	Fine-loamy, siliceous, thermic Aquic Paleudults
Kilmanagh	Fine-loamy, mixed, nonacid, mesic Aeric Haplaquepts
Lochloosa	Loamy, siliceous, hyperthermic Aquic Arenic Paleudults
Londo	Fine-loamy, mixed, mesic Aeric Glossaqualfs
Lufkin	Fine, montmorillonitic, thermic Vertic Albaqualfs
Malbis	Fine-loamy, siliceous, thermic Plinthic Paleudults
Mantachie	Fine-loamy, siliceous, acid, thermic Aeric Fluvaquents
McCook	Coarse-silty, mixed, mesic Fluventic Haplustolls
Millhopper	Loamy, siliceous, hyperthermic Grossarenic Paleudults
Mollville	Fine-loamy, siliceous, thermic Typic Glossaqualfs
Munjor	Coarse-loamy, mixed (calcareous), mesic Typic Ustifluvents
Myakka	Sandy, siliceous, hyperthermic Aeric Haplaquods
Newco	Clayey, mixed, thermic Aquic Hapludults
Parkhill	Fine-loamy, mixed, nonacid, mesic Mollic Haplaquepts
Penden	Fine-loamy, mixed, mesic Typic Calciustolls
Pipestone	Sandy, mixed, mesic Entic Haplaquods
Pleasant	Fine, montmorillonitic, mesic Torrertic Argiustolls
Poarch	Coarse-loamy, siliceous, thermic Plinthic Paleudults

(continued on next page)

Series	Family
Pomello	Sandy, siliceous, hyperthermic Arenic Haplohumods
Rapson	Sandy over loamy, mixed, mesic Entic Haplaquods
Riverdale	Loamy, mixed, mesic Aquic Arenic Hapludalfs
Samsula	Sandy or sandy-skeletal, siliceous, dysic, hyperthermic Terric Medisaprists
Sanilac	Coarse-silty, mixed (calcareous), mesic Aeric Haplaquepts
Shebeon	Fine-loamy, mixed, mesic Aeric Ochraqualfs
Smyrna	Sandy, siliceous, hyperthermic Aeric Haplaquods
St. Johns	Sandy, siliceous, hyperthermic Typic Haplaquods
Tavares	Hyperthermic, uncoated Typic Quartzipsamments
Tobico	Mixed, mesic Mollic Psammaquents
Troup	Loamy, siliceous, thermic Grossarenic Kandiudults
Urland	Clayey, mixed, thermic Typic Hapludults
Wabasso	Sandy, siliceous, hyperthermic Alfic Haplaquods
Wasepi	Coarse-loamy, mixed, mesic Aquollic Hapludalfs
Yonges	Fine-loamy, mixed, thermic Typic Ochraqualfs

REFERENCES

Bouma, J. 1986. Using soil survey information to characterize the soil-water state. J. Soil Sci. 37:1–7.

Brown, R.B. 1988a. Concerning the quality of soil survey. J. Soil Water Conserv. 43:452–455.

Brown, R.B. 1988b. Relevant interpretations for the 1990's. p. 65–86. In H.R. Finney (ed.) Soil resources: Their inventory, analysis, and interpretation for use in the 1990's. Proc. Int. Interactive Workshop, Minneapolis, MN. 22–24 Mar. 1988. Educational Development System, Minnesota Extension Service, Minneapolis, MN.

Brown, R.B., and G.A. Miller. 1989. Extending the use of soil survey information. J. Agron. Educ. 18:32–36.

Devore, J.L., and R.L. Peck. 1986. Statistics, the exploration and analysis of data. West Publ. Co., St. Paul, MN.

Di, H.J., B.B. Trangmar, and R.A. Kemp. 1989. Use of geostatistics in designing sampling strategies for soil survey. Soil Sci. Soc. Am. J. 53:1163–1167.

Doolittle, J.A., G. Schellentrager, and S. Ploetz. 1989. Soil survey of Hillsborough County, Florida. USDA-SCS in cooperation with Univ. Florida IFAS, Agric. Exp. Stn. and Soil Sci. Dep., and Florida Dep. Agric. and Consumer Services. U.S. Gov. Print. Office, Washington, DC.

Fox, B.C. 1989. Soil survey of Conecuh County, Alabama. USDA-SCS in cooperation with Alabama Agric. Exp. Stn. and Alabama Soil and Water Conserv. Comm. U.S. Gov. Print. Office, Washington, DC.

Hamilton, V.L., R.C. Angell, and B.D. Tricks. 1989. Soil survey of Decatur County, Kansas. USDA-SCS in cooperation with Kansas Agric. Exp. Stn. U.S. Gov. Print. Office, Washington, DC.

Hendrickx, J.M.H., L.W. Dekker, M.H. Bannink, and H.C. van Ommen. 1988. Significance of soil survey for agrohydrological [sic] studies. Agric. Water Manage. 14:195–208.

Horvath, E.H., E.A. Fosnight, A.A. Klingebiel, D.G. Moore, and J.E. Stone. 1987. Using a spatial and tabular database to generate statistics from terrain and spectral data for soil surveys. p. 91–98. In W.U. Reybold and G.W. Petersen (ed.) Soil survey techniques. SSSA Spec. Publ. 20. SSSA, Madison, WI.

Linsemier, L.H. 1980. Soil survey of Huron County, Michigan. USDA-SCS in cooperation with Michigan Agric. Exp. Stn. U.S. Gov. Print. Office, Washington, DC.

Miller, A. 1983. Integrated pest management: Psychosocial constraints. Prot. Ecol. 5:253–267.

Miller, F.P., and R.B. Brown. 1987. Future developments in the private sector related to soil genesis, morphology and classification. p. 269–278. In L.L. Boersma et al. (ed.) Future developments in soil science research. SSSA, Madison, WI.

Mowery, I.C., H. Oakes, J.D. Rourke, F. Matanzo, H.L. Hill, G.S. McKee, and B.B. Crozier. 1958. Soil survey of Brazos County, Texas. USDA-SCS in cooperation with Texas Agric. Exp. Stn. U.S. Gov. Print. Office, Washington, DC.

Neitsch, C.L. 1982. Soil survey of Jasper and Newton Counties, Texas. USDA-SCS and USDA-FS in Cooperation with Texas Agric. Exp. Stn. U.S. Gov. Print. Office, Washington, DC.

Ostle, B., and L.C. Malone. 1988. Statistics in research: Basic concepts and techniques for research workers. 4th ed. Iowa State Univ. Press, Ames, IA.

Robert, P.C., and J.L. Anderson. 1987. A convenient soil survey information system (SSIS). Appl. Agric. Res. 2:252-259.

Schellentrager, G.W., J.A. Doolittle, T.E. Calhoun, and C.A. Wettstein. 1988. Using ground-penetrating radar to update soil survey information. Soil Sci. Soc. Am. J. 52:746-752.

Snedecor, G.W., and W.G. Cochran. 1980. Statistical methods. 7th ed. Iowa State Univ. Press, Ames, IA.

Soil Survey Staff. 1983. National soils handbook. USDA-SCS 430-VI-NSH. U.S. Gov. Print. Office, Washington, DC.

Stein, A., M. Hooderwerf, and J. Bouma. 1988. Use of soil-map delineations to improve (co-)kriging of point data on moisture deficits. Geoderma 43:163-177.

U.S. Department of Agriculture-Soil Conservation Service. 1990. Soil series of the United States, including Puerto Rico and the U.S. Virgin Islands. Misc. Publ. 1483. U.S. Govt. Print. Office, Washington, DC.

Wilding, L.P. 1988. Improving our understanding of the composition of the soil-landscape. p. 13-39. *In* H.R. Finney (ed.) Soil resources: Their inventory, analysis, and interpretation for use in the 1990's. Proc. Int. Interactive Workshop, Minneapolis, MN. 22-24 Mar. 1988. Educ. Devel. System, Minnesota Extension Service, Minneapolis, MN.

Wonnacott, R.J., and T.H. Wonnacott. 1985. Introductory statistics. 4th ed. John Wiley & Sons, Inc., New York.

9 Soil Mapping Concepts for Environmental Assessment

Duane A. Lammers

USDA-FS
USEPA Environmental Research Laboratory
Corvallis, Oregon

Mark G. Johnson

METI, Inc.
USEPA Environmental Research Laboratory
Corvallis, Oregon

ABSTRACT

Spatial variability of soils in the landscape and how this variability is represented by soil maps and portrayed in map unit descriptions are critical for assessing many of today's pressing environmental concerns. Traditional concepts used in mapping, naming, defining, and correlating map units need to be redirected from a focus on taxonomic or use criteria to the use of soil properties as the basis for correlation. Map unit delineations need to represent real segments of the landscape by capturing discrete soil patterns that function as workable stratifications of the landscape and provide detailed soil information for use in environmental assessment. A reliable estimate of the proportionate extent of map unit components is needed. By grouping soils into response classes, minor soil components that are important to understanding landscape dynamics and assessing environmental concerns can be retained in regional-scale databases. Current examples of regional and global soil databases are presented and a global framework approach for nesting mapping databases of different scales is proposed.

Scientists from many disciplines are recognizing the vital role that soils play in mitigating the effects of natural and anthropogenic perturbations (e.g., fire, acid rain, global climate change) of ecosystems. Because many environmental problems are regional, continental, or global in scale, an alternate strategy for developing soil databases that capture local-scale soil spatial variability and provide a mechanism for maintaining integrity across scales of extrapolation needs to be developed.

In this chapter, we explore an approach to linking soil mapping data needed for local-scale process studies to global-scale databases. Changes in traditional concepts used in the design of map units and interpretation of soils are recommended to facilitate linking of databases from one scale to another. Current examples of regional and global soil databases are presented and a global framework approach for nesting mapping databases of different scales is proposed.

A general hypothesis for using soils in environmental assessments can be stated as follows: the kinds of soil and their distribution in the landscape affect the response of ecosystems to environmental stress. We assume that soil properties in concert with other ecosystem attributes (i.e., climate, topography, biota, and lithology) are related to identifiable landscape segments that can be mapped. Therefore, if we can adequately map and characterize the soils and develop models that simulate soil processes for individual soils or classes of soils; and, if we can aggregate effects for landscape units and extrapolate results to regions, we can project the effect of a stressor in a region. Soil processes are influenced by both internal properties and by characteristics of soils in neighboring landscape segments. Consequently, our understanding of relationships among soils and environmental effects is limited by pedologists' ability to capture data about soils and their distribution in a landscape at a scale appropriate for linking data to process models.

CONCERNS WITH TRADITIONAL SOIL SURVEYS

Traditional concepts that have evolved with soil mapping need to be examined as a basis from which to propose changes that will facilitate adaptation of soils information to geographic information systems, electronic databases, and global assessments. Products of the National Cooperative Soil Survey (NCSS) in the USA have evolved around interpretations needed for farm planning and conservation of the agricultural soil resource. Purity standards placed on soil mapping and on the extent of similar and dissimilar components of map units (e.g., a consociation must consist of 75% or more of the taxon for which the map unit is named and similar taxon) (Soil Survey Staff, 1983) were a reflection of the expected use of the maps and to a degree have misrepresented the true map unit composition. Map unit purity standards reviewed in Buol et al. (1980) and Wilding (1988) reported purity levels of about 50% for soil map units named for soil series. Emphasis, in the past, has been on interpretative units rather than naturally occurring delineations of the landscape, resulting in map units focused on interpretation for a specific purpose. Although satisfactory for the specific purpose for which the mapping was done, this approach may not have produced the best soil database for other applications.

Soil taxonomy has provided a system to classify soils described in the mapping process, but has also been used as a correlation "shoehorn" to force decisions on which soils were similar or dissimilar. This comprehensive classification system has focused attention on class boundaries and has undoubtedly biased the selection of modal representatives of a class. A soil became "different" if one or more properties fell outside the limits of family criteria. Nettleton and Brasher (1989) have proposed a solution to this "taxadjunct problem" that has resulted from application of rigid taxonomic limits to soil series and other "taxonomic units." Soil taxonomy provided a framework for relationships among soils, and among soils and the factors responsible for their character. Taxonomic classification does not usually form soils into the best groups for making predictions for a particular purpose. Although taxonomic classes help us to recognize fundamental differences in soils, it is often necessary to regroup soils into classes that relate to specific problems (Soil Survey Staff, 1975). For example, if soil wetness is thought to be an important property for an assessment, the 11 soil orders are not appropriate classes. Soils would need to be grouped by depth-to-water table, drainage, landscape position or other soil-wetness related characteristics.

HOW ARE SOIL SURVEYS USED IN ENVIRONMENTAL ASSESSMENTS?

Environmental assessments are often used to make regional predictions that answer a variety of "what if" questions concerning stress on terrestrial ecosystems. Soils are integral in the natural mitigation of ecosystem perturbations and are being included more frequently in the assessment of the extent of environmental damage or change. Soils within a region of environmental concern can be inventoried and characterized; and simulation models can be developed that use the soils data to predict the response of watersheds or other ecosystems in the region. The specific location of soils within a landscape segment may not always be needed to determine the cumulative response for that segment of the landscape. Alternatively, by combining the effects of individual soils or groups of soils with their extent, the region can be characterized and cumulative effects on the region determined. Spatial relationships among components of map units and among map units that comprise landscape systems, however, are useful in describing system dynamics. The soil database structure and data aggregation procedure(s) for an assessment will depend on the importance of specific map unit components and spatial distribution of those components.

WHAT MAP UNIT CHARACTERISTICS ARE NEEDED?

Soil map units need to be synchronized with other landscape attributes (e.g., land use, vegetation, climate) so relationships that naturally occur in the landscape are portrayed on maps. In other words, the portrayal of relational landscape attributes must be synthesized to produce correlation of data layers in geographic information systems. Failure to properly correlate the attributes will result in discordant landscape segments where unrelated types of attributes intersect.

Ideally, map units should represent delineation of landscape segments for which natural processes and spatial variability of those processes cause an individual segment to be different from neighboring segments. The kinds and extent of soils in a region need to be determined from soil map units delineated on a map and described in terms of their composition. The proportion of map unit components must be quantified at a level of reliability required by the data quality needs of assessment models. If minor components of map units prove to be important to modeling landscape processes and assessing environmental concerns, they must be retained in regional-scale databases. For example, if we assume that narrow (30–60 m) riparian or near-riparian areas adjacent to surface waters are critical as "scrubber" zones for nonpoint pollutants, their presence in an agroecosystem landscape may have a much larger effect on surface water quality than their proportionate extent in the landscape. These minor, but critical, components are usually either lost in the "generalization" process, or are not found by small-scale remotely sensed mapping that may be used in the development of regional-scale soil maps and databases.

WHAT ARE THE NEEDS FOR SOIL DATA?

In the conduct of regional-scale environmental assessments, processes that realistically apply to individual soil pedons or to groups of soils with selected characteristics in common need to be extrapolated to the region. Differences in properties from one layer of the soil to another may have an important effect on process model predictions. If models are sensitive to "within" soil variability, then data for individual pedon horizons or for major horizons representing a group of pedons in the same class may be needed. Feedback can be provided by testing of models and evaluation of soil properties that are important to segregating soils into classes. The important properties will probably be different for the different environmental concerns that might be assessed. Likewise, soil properties that relate well to ecosystem characteristics in one region may be insignificant in another region. For example, texture of the surface horizon may discriminate organic C content on a loess-covered prairie, but not in a glaciated upland. Soil properties, determined to be important for use in addressing a broad range of environmental concerns, need to be identified for retention in a universal regional or global-scale soil database.

Spatial variability of soil properties within a region is usually much too complex to use properties of every contrasting soil horizon for a large number of pedons in model calculations with the results summed across the region. Computing power needed to simultaneously run simulation models at this level of spatial resolution for a large region is not practical. Conversely, if models use mean values for the region, an erroneous result may be obtained because relationships are not normally distributed.

A system of related data sets that facilitate retention of important characteristics while moving from one scale to the next is needed. As pedologists, we must develop realistic approaches that deal with this across-scale issue. We need to quantify the spatial variability of soil components within map units at a given scale and seek ways to aggregate soil data in such a manner that the integrity of important soil properties is preserved.

APPROACHES TO GLOBAL SOIL DATABASES

For the most part, "currently available" or "in progress" regional- or global-scale soil databases stand alone with respect to map unit legend, map unit composition, and relationship to other scales of mapping. Bowman (1989) has recently reviewed global soil maps and data sets, and summarized their usefulness and limitations for use in global environmental assessments. The following soil databases and their respective approaches are presented as examples of soil data at different mapping scales and are discussed in terms of their potential use for regional or global assessments. Linkages between these databases or other similar databases need to be constructed to enable their use in the transfer of local-scale process results to global-scale systems. These soil databases need to be fully evaluated for application and quantitative reliability in assessment of environmental concerns.

Food and Agricultural Organization of the United Nations (1:5 000 000)

The FAO Soil Map of the World (Food and Agriculture Organization/United Nations Educational, Scientific, and Cultural Organization, 1971–1981) is a classification of map units that can be used to reflect general soil patterns in a large region. This database does not provide information on soil properties of individual pedons or classes of soils as needed for most process models. Furthermore, soil spatial variability depicted by maps at this scale would be inadequate for simulation of landscape processes.

Soil and Terrain Digital Database (1:1 000 000)

Compilation of soil information has recently started in pilot areas using procedures developed for a World Soil and Terrain (SOTER) Digital Database. The SOTER data base uses a universal legend based on soil properties and is designed to accommodate all national classification systems. Separate attribute files are used to record characteristics of map delineations, terrain components and soil layers. Each map unit is described by one to three terrain components and represented by one to three soils. The proportionate extent of each terrain component in a map unit is recorded in the terrain attribute file. Where there are existing soil maps, SOTER map units are generalizations of source map units (Shields & Coote, 1989). Although local-scale soil maps, where available, are used to construct the SOTER map, there is no apparent attempt to link data on map units or map unit components between scales. The SOTER attribute files provide both data on soil properties and estimates of the proportion of kinds of soils in map units. Terrain components and soils of lesser extent fall through the cracks of the system and are lost from the database. Completion of global coverage with the SOTER is expected to require about 20 yr.

State Soil Geographic Data Base (1:250 000)

Under the National Cooperative Soil Survey in the USA, the SCS is compiling a regional-scale State Soil Geographic Data Base (STATSGO) from existing detailed soil maps (Reybold & TeSelle, 1989; Bliss & Reybold, 1989). As many as 21 components and their estimated proportions in the map unit are allowed. Map unit components, mostly phases of soil series, are linked through a map unit use file to appropriate soil interpretations records. In this manner, soil map units can be linked to soil properties needed to support process models. A link has not been established that will allow us to zoom in for a closer look at the more detailed delineations that make up an individual map unit of the STATSGO map. Map unit composition was based on "desk top" transects of existing large-scale soil maps where they were available and relied predominately on aerial photography interpretation of areas not previously mapped. An element of ground truth was included in the mapping of the existing maps. With this approach, map unit composition is based on named components of the existing map units, but, excludes their minor components. Because traditional soil surveys are the basis for the STATSGO maps, the concerns discussed in the section above are applicable. Transects could be conducted in randomly selected delineations of the STATSGO map to evaluate map unit composition and to identify components that may not have been included.

CHARACTERISTICS NEEDED FOR NESTED SOIL DATABASES

Nesting of maps and their associated databases was perceived as a method for linking spatially distributed terrestrial attributes of different scales. It is a system that will allow us to "zoom in" on local-scale maps for a closer look at the heterogeneity of individual regional-scale map units and "zoom out" to regional scale with important data or results from soil processes. Characteristics needed for nesting soil databases include a global skeleton or framework, delineation of naturally occurring segments of the landscape, and quantification of map units and retention of soil properties in the database.

Framework

A geographic framework is needed to facilitate understanding and characterization of global ecosystem patterns and their components. In addition to providing a mechanism for determining within-region variability and a means to measure representativeness of sites within regions, a framework is needed to maintain integrity of attribute composition when moving from a map of one scale to another. A framework provides a means to ensure spatial congruence of data. If, for example, we attempt to overlay a soil map and a vegetation map of a region that were developed separately and without synthesis of attributes and congruence of map delineations, several more soil-vegetation combinations may be created than actually exist. These incongruent data intersections tend to mask true soil-vegetation relationships and lessen the precision of our predictions.

The geographic framework constructed by Omernik (1987) for delineating ecoregions of the USA, depicts terrestrial ecosystems by conceptualizing distinct patterns of terrestrial characteristics. In the development of ecoregions for the USA, Omernik (1987) hypothesized that ecosystems and their components display regional patterns that are reflected in spatially variable combinations of attributes including soils, geology, climate, vegetation, physiography and land use. This concept of an ecoregion embodies the synthesis of these attributes to discern individual regions. An ecoregion is subtly different from other regional classifications that focus on differences in just one or two dominant attributes to characterize regions.

Map Units Delineated on Natural Landscape Units

Map units designed to represent naturally recurring segments of the landscape will allow linking of spatially distributed terrestrial attributes at different scales of mapping. Alternatively, map units that are designed for a specific use or interpretation, or delineated along artificial boundaries, will cross natural landscape breaks and, therefore, not permit nesting. For environmental assessment, map units must convey the relationship between soil distribution and the landscape at a scale that captures important soils. We cannot expect to relate process models of ecosystem disturbance to the landscape if we fail to delineate the spatial units of our database in a manner that allows us to discern differences in the natural landscape processes.

Quantification of Map Unit Components

The composition of each map unit needs to be rigorously quantified for reliable data aggregation, linkage between scales, and regionalization. Quantification can only progress from fine to coarse resolution; therefore, map unit composition must be determined for the largest-scale (finest-resolution) map of those being nested. Individual delineations of map units can be randomly selected for transecting, intense mapping, or grid-point examination to obtain data from which we can make a reliable quantitative estimate of the proportionate extent of map unit components. Map unit composition can be quantified at any map scale, but must be consistent for nesting map units of different scales and have the flexibility to be updated as new data become available.

Soil Properties in Database

Soil correlation is a decision process by which soils with similar characteristics are grouped together. Soils that respond similarly to a specific use are often thought of as "similar" soils. Soils may also be grouped by genetic classification or according to a similar set of properties. Taxonomic or interpretive groupings often result in a high degree of intraclass variability of basic soil properties. Soil properties that have been described in the field and measured in the laboratory need to be retained in a database and used as the basis for segregating soil classes (e.g., soil series). Soil properties that influence productivity of vegetation, biogeochemistry, and hydrology are important in environmental assessment. If soils are correlated on important properties with respect to environmental concerns, they can later be grouped for a specific concern to reduce database size and improve modeling efficiency.

Example from the Direct/Delayed Response Project

A project designed to predict the future effects of acidic deposition on soils and surface waters, the Direct Delayed Response Project (DDRP) used soil mapping and sampling of soil groups to characterize soil properties within regions (Lee et al., 1989a). Within a defined region, a statistically representative set of watersheds was selected for mapping and characterization of soils and other watershed attributes. This probabilistic structure provided the basis for extrapolation of measured and modeled results from the watersheds to the region without surveying the entire region. Data collected along randomly selected transects were used to make "best estimates" of the composition of map units for which the proportion of each component was recorded to the nearest 5%. Map unit components (mostly phases of soil series) were grouped into classes based on soil properties considered to be important in the assessment of the effects of acidic deposition on surface water chemistry. These classes were characterized by sampling and laboratory analyses of six to eight pedons representing each class. This approach to a regional database provided data on soil properties required for dynamic watershed simulation models (Church, 1989), as well as soil distribution in the landscape at an appropriate scale. Map unit components were linked to important soil characteristics and taxonomic classification, and could be easily linked to appropriate soil interpretations records (Lee et al., 1989b).

An example from the DDRP Southern Blue Ridge region is used to demonstrate the linking of nested databases. Watersheds were selected within a regional framework; naturally recurring segments of the watersheds were delineated on maps; the map unit composition was quantified from transects and other field observations; and soil properties considered most likely to influence surface water chemistry in the region were used to group map unit components into classes. Watersheds were mapped at a scale of 1:24 000 (the map in Fig. 9-1 was printed at a scale of 1:48 000 for this publication).

Although soils were correlated to one legend within a region, we will discuss just the portion of the regional legend applicable to the watershed

SOIL MAPPING FOR ENVIRONMENTAL ASSESSMENT

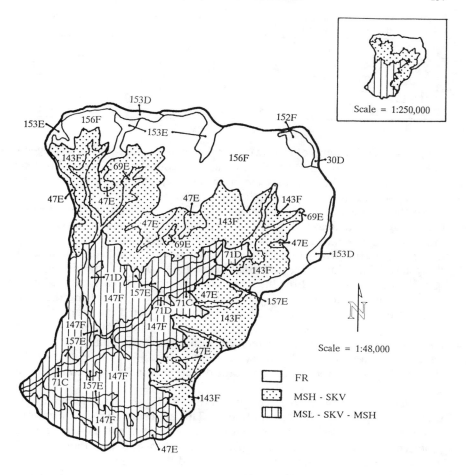

30D	Craggy-Rock outcrop complex, 15-40% slopes
47E	Jeffrey cobbly loam, 25-50% slopes
69E	Spivey cobbly loam, 30-50% slopes
71C	Spivey-Santeetlah complex, 8-15% slopes
71D	Spivey-Santeetlah complex, 15-30% slopes
143F	Jeffrey-Spivey cobbly loams, 50-90% slopes
147F	Unicoi-Ditney-Rock outcrop complex, 50-90% slopes
152F	Spivey Variant cobbly loam, 50-90% slopes
153D	Burton Taxadjunct cobbly loam, 15-30% slopes
153E	Burton Taxadjunct cobbly loam, 30-50% slopes
156F	Umbric Dystrochrepts-Craggy-Rock outcrop complex, 50-90% slopes
157E	Brookshire loam, 30-50% slopes

Fig. 9-1. Soil map of a watershed depicting: (i) map units (identified by map symbols) consisting of components that are mostly phases of soil series at a scale of 1:48 000 and (ii) map units (identified by shaded areas) that are comprised of several of the smaller delineations and have map components consisting of soil classes that are groups of the phases of soil series map unit components. The larger delineations are shown at a "regional," 1:25 000 scale on the inset map.

Table 9-1. Proportionate areal extent (%) of soil classes of the three map units represented by shaded areas in Fig. 9-1.

Map unit	Soil Class						
	FR	MSH	MSL	MRO	SKV	SKX	OTL
FR	91		<1	8			
MSH-SKV		58	1		40	<1	
MSL-SKX-MSH		18	33	11	8	29	1

in Fig. 9-1 in this example. Twelve different map units (listed in the soil legend in Fig. 9-1) composed of 40 different map unit components were delineated on the watershed. The procedure used to group map unit components resulted in 12 classes for the region (Church et al., 1989); components that represent six of the classes occur on this watershed. Three "generalized" map units (shown by shaded areas in Fig. 9-1) were delineated by combining associated map units that occur in a repeating pattern on the landscape and represent differences in the proportion of the six soil classes. Once the boundaries of these new delineations were identified in the geographic information system (GIS) data file, we could readily determine the extent of each of the large-scale map units within each of the three small-scale units. Using the map unit composition file with the GIS data set, we calculated the proportionate extent of the classes in each of the small-scale map units (Table 9-1). The result is a soil map of the watershed with three map units that can be delineated at a scale of 1:250 000 (see insert in Fig. 9-1). The extent of the six soil classes that represent important differences in soils is retained in a database.

The extent of field work required in the DDRP approach may limit its usefulness for large regions or continents. This approach to regionalization, however, may be very useful in characterizing regions and in determining representativeness of study sites used to develop and test models. The quantitative nature of the DDRP database and the procedure used to group soils demonstrates a realistic approach that could be used in linking soil databases.

CONCLUSIONS

The need for soils data to assess environmental concerns on regional and global scales demands the development of flexible, multipurpose, and scale transferrable databases. We can no longer think of soil surveys as being conducted for a specific purpose and soil map units as delineations of the landscape having a singular response to use and management. Map units must be aligned with recognizable repeating segments of the landscape, which are correlated on the basis of differences in the proportion of map unit components.

Components of map units must represent actual differences in soil properties, not differences in taxonomic class or response to a specific use.

These soil components can then be grouped into classes to facilitate the most precise predictions for a specific use or concern. The relative proportion of map unit components needs to be determined quantitatively and landscape characteristics of their occurrence described. Some thought must be given to defining separate map units on the basis of kinds and extent of contrasting inclusions.

Procedures need to be developed for "nesting" levels of mapping in a way that will allow transfer of spatial distribution, and physical and chemical properties of soils that are applicable to soil processes from one scale to another. Through careful description and quantification of the soil map unit, spatial variability can be linked to characteristics of the map unit components. Field and laboratory data of individual pedons or groups of similar pedons can be used to characterize the kinds of soil (e.g., soil series) in the landscape. These data can be input to process models to simulate effects of stressors on the environment.

By developing an ecoregion-type framework for the orderly nesting of soils databases and by employing procedures for aggregation of data, modeling of systems and extrapolation of results, individual processes can be scaled-up to regions or continents. The occurrence of small areas of contrasting soil may be of great importance in explaining the hydrology, habitat, or buffering capacity of the landscape. When aggregating up to a smaller-scale map base, these key segments of the landscape must not be lost or distorted in the process. The extent of soil bodies and their position in relation to other soil bodies and to surface waters in the landscape are important to understanding and assessing effects on terrestrial systems.

REFERENCES

Bliss, N.B., and W.U. Reybold. 1989. Small-scale digital soil maps for interpreting natural resources. J. Soil Water Conserv. 44:30–34.

Bowman, A.F. 1990. Global distribution of the major soils of the world and land cover types. p. 33–59. *In* A.F. Bowman (ed.) Soils and the greenhouse effect. John Wiley & Sons, New York.

Buol, S.W., F.D. Hole, and R.J. McCracken. 1980. Soil genesis and classification. 2nd ed. Iowa State Univ. Press, Ames, IA.

Church, M.R. 1989. Direct/delayed response project: Predicting future long-term effects of acidic deposition on surface water chemistry. EOS Trans. Am. Geophys. Union 70:801–03, 812–13.

Church, M.R., K.W. Thornton, P.W. Shaffer, D.L. Stevens, B.P. Rochelle, G.R. Holdren, M.G. Johnson, J.J. Lee, R.S. Turner, D.L. Cassell, D.A. Lammers, W.G. Campbell, C.I. Liff, C.C. Brandt, L.H. Liegel, G.D. Bishop, D.C. Mortenson, S.M. Pierson, and D.D. Schmoyer. 1989. Direct/delayed response project: Future effects of long-term sulfur deposition on surface water chemistry in the Northeast and Southern Blue Ridge Province. EPA/600/3-89/061a-d. U.S. EPA, Washington, DC.

Food and Agriculture Organization/United Nations Educational, Scientific, and Cultural Organization. 1971 to 1981. Soil map of the world, 1:5 000 000. Vol. I to IX. FAO, Rome.

Lee, J.J., D.A. Lammers, M.G. Johnson, M.R. Church, D.L. Stevens, D.S. Coffey, R.S. Turner, L.J. Blume, L.H. Liegel, and G.R. Holdren. 1989a. Watershed surveys to support an assessment of the regional effects of acidic deposition on surface water chemistry. Environ. Manage. 13:95–108.

Lee, J.J., D.A. Lammers, D.L. Stevens, K.W. Thornton, and K.A. Wheeler. 1989b. Classifying soils for acidic deposition aquatic effects: A scheme for the northeast USA. Soil Sci. Soc. Am J. 53:1153–1162.

Nettleton, W.D., and B.R. Brasher. 1989. The taxadjunct problem. p. 269. *In* Agronomy abstracts. ASA, Madison, WI.

Omernik, J.M. 1987. Ecoregions of the coterminous United States. Annu. Assoc. Am. Geogr. 77:118–125.

Reybold, W.U., and G.W. TeSelle. 1989. Soil geographic data bases. J. Soil Water Conserv. 44:28–29.

Shields, J.A., and D.R. Coote. 1989. SOTER procedures manual for small scale map and data base compilation and procedures for interpretation of soil degradation status and risk (for discussion). ISRIC Working Pap. and Preprint 88/2, rev. ISRIC, Wageningen, Netherlands.

Soil Survey Staff. 1975. Soil taxonomy: A basic system of soil classification for making and interpreting soil surveys. USDA-SCS Agric. Handb. 436. U.S. Gov. Print. Office, Washington, DC.

Soil Survey Staff. 1983. National soils handbook. USDA-SCS Agric. Handb. 430. U.S. Gov. Print. Office, Washington, DC.

Wilding, L.P. 1988. Improving our understanding of the composition of the soil landscape. p. 13–35. *In* M.R. Finney (ed.) Proc. Int. Interactive Workshop on Soil Resources: Their inventory, analysis and interpretation for use in the 1990's. Minnesota Ext. Serv., Univ. Minnesota, St. Paul, MN.

10 Minimum Data Sets for Use of Soil Survey Information in Soil Interpretive Models

R. J. Wagenet
Cornell University
Ithaca, New York

J. Bouma
Agricultural University
Wageningen, Netherlands

R. B. Grossman
Soil Conservation Service
Lincoln, Nebraska

ABSTRACT

Simulation modeling is increasing in importance as a tool for soil survey interpretations. Simultaneous consideration of both agronomic and environmental interpretations will be required as soil management systems are designed that allow optimal, sustainable production while minimizing adverse environmental consequences. Taxonomic data, as currently obtained in soil survey, can be used to estimate basic parameters needed for simulation by defining pedotransfer functions that relate simple soil characteristics to more complicated model parameters. Some important data are not currently being gathered during soil survey, such as properties of the soil surface, measurement of soil bulk densities, and temporal (seasonal) changes in properties. Satisfactory application of models is possible only if such data are obtained in the future.

Soil survey activities have traditionally focused on the measurement and classification of soil and land characteristics and properties. As defined by the Food and Agriculture Organization (1976), land characteristics are attributes of land that can be measured or estimated, usually during soil surveys (e.g., soil texture, structure, organic matter content, slope angle). Some characteristics are increasingly needed for modern land evaluation; but

---

Copyright © 1991 Soil Science Society of America, 677 S. Segoe Rd., Madison, WI 53711, USA. *Spatial Variabilities of Soils and Landforms.* SSSA Special Publication no. 28.

they are not routinely determined by soil survey (e.g., bulk density, moisture retention, hydraulic conductivity, phosphate adsorption, cation exchange capacity). Additionally, distinction was suggested by Bouma and van Lanen (1987) between basic characteristics that can be measured in the laboratory or estimated in the field (e.g., soil texture, organic matter content, Fe and Al content) and other characteristics (called soil properties in the abovementioned paper) that can be estimated from basic soil characteristics (e.g., bulk density estimated from texture, moisture retention estimated from texture and bulk density, cation exchange capacity estimated from texture and organic matter content, and phosphate adsorption estimated from Fe and Al content). Traditional soil survey activities, now nearing completion in many areas of the USA and the developed world, have measured, categorized, tabulated, subdivided and to a much lesser degree interpreted land characteristics.

As societal concerns about land-use planning and environmental pollution have intensified, the soil scientist is increasingly being asked to play a central role in solving the problems that arise, for many of the questions facing society today require an assessment of the dynamic behavior of soils. These questions may focus on the management of a particular combination of soil, crop, climate and irrigation, the use of a particular soil as a waste disposal site or the potential that a particular soil–water–chemical system represents with respect to groundwater pollution. Although the static attributes of soil characteristics and properties are useful, beginning points in answering the questions, they provide only partial answers since they tell us little about soil processes that are use-dependent, and therefore temporally variable. A useful, even necessary next step is to establish land qualities, which are defined by the Food and Agriculture Organization (1976) as complex attributes of land that act in a distinct manner to influence the suitability of land for a specific kind of use. The definition implies that land qualities cannot directly be measured (e.g., water supply capacity, trafficability, temperature regime, erosion hazard). Land qualities can be related to land characteristics through a variety of methods.

Two of the most important methods are pedotransfer functions and simulation modeling. The former serves to translate, through empirical, regression or functional relationships, the basic information found in the soil survey into a form useful in broader applications, such as simulation modeling. Examples of pedotransfer functions are expressions relating soil texture and organic matter to the water flow properties of a soil, or a statistically derived relationship between phosphate sorption and the Al or Fe content of a soil. Such attempts to generalize the standard characterization data in terms of numerical relationships are being increasingly pursued in both the USA and Western Europe. However, pedotransfer functions only provide a part of the solution to estimating land qualities, as they are essentially filters through which basic soil survey information passes to eventually be used in simulation models. It is through the use of such models that intrinsic soil attributes are converted into, or interpreted in terms of, dynamic land qualities. A number of models, such as WEPP, DRAINMOD and LEACHM (all discussed below), currently make routine use of formal relationships between standard characterization data and more basic properties, and use these to predict soil processes over depth and time. Such a process of using basic information to derive the suitability of land for a particular use is called land evaluation, which according to the original definition of the Food and Agriculture Organization (1976) is concerned with estimating land performance when the land is used for specific purposes.

Simulation modeling is used increasingly to address problems of land evaluation and environmental protection. A number of models now exist that can be applied to questions of water infiltration and runoff, crop growth and extraction, and chemical leaching and groundwater contamination. These models vary in complexity, and in the subset of all possible processes that they consider. Identification of basic soil and land characteristics and properties should be considered a first step in using these models. This may necessitate some change in standard soil survey procedures. For example, soil survey in the USA has been classically concerned with the preparation of soil maps that would have application over periods of time measured in decades. For this reason, the emphasis has been placed on relatively permanent properties that would not greatly change over time. In addition, the management of soil survey has dictated that a relatively small amount of time (hence, money) would be spent per unit area. The average amount of field time in the U.S. soil survey is approximately hundreds per hectare. This small amount of time per unit area necessitates that the delimited regions commonly have a large variability in characteristics important to use. The soil map unit concept in the U.S. soil survey is supposed to describe the various kinds of soils and their areal disposition with emphasis on management effects. These map unit descriptions commonly do not supply the detail necessary to understand soil behavior in the use-specific, temporal sense so important in simulation modeling. The concepts are particularly weak in the information pertaining to both overland flow of water and leaching of solutes through the unsaturated zone.

There is an alternative approach that will enhance the opportunity for soil survey information to be used for simulation modeling. Small representative areas could be mapped at sufficient detail to apply the current taxonomic criteria. These areas could be published as illustrative of standard map unit concepts. Relatedly, soil survey personnel could collect field observations and measurements pertaining to use-specific, temporal soil properties as part of standard soil survey operations. Such changes should be strongly considered in the future.

THE DATA DEMANDS OF SIMULATION MODELS

Models that simulate the dynamic behavior of soils generally presume that the system operates such that the occurrence of a given set of events leads to a uniquely definable consequence. Such models are termed deterministic models, and are the type of models currently used in the land evaluation process. In order to be executed for predictive purposes, they require input information about soil characteristics and properties, as well as information about the conditions of water application. They also require as input certain initial conditions of the soil profile (defined below) at the beginning of the time period to be simulated. The exact nature of this required information, particularly for water flow properties, depends upon which general type of deterministic model is used.

These models are generally separated by the manner in which they represent basic processes of infiltration, water flow and chemical transport. At one extreme are mechanistic models, which have been developed on the recognition that potential energy gradients are the driving force for water flow. A hydraulic conductivity is defined to relate the water flux and hydraulic gradient. This model is termed *mechanistic* as it is presumed to be representative of our best current understanding of basic mechanisms of water flow.

At the opposite extreme are functional models, which greatly simplify physical process in order to reduce data demands and computational time. Intermediate cases of pseudomechanistic models also exist, in which a compromise is sought between representation of basic process and the operational realities of data collection capabilities and required accuracy of model output. Examination of several models will illustrate the differences in both structure and data demands.

The common mechanistic approach for describing water flow through unsaturated soil is Darcy's law, which when applied to transient cases of water movement in a one-dimensional system becomes the Richard's equation

$$\partial\theta/\partial t = \partial/\partial z \, [K(\theta) \, \partial H/\partial z] \qquad [1]$$

where θ is volume water content; $\partial H/\partial z$ is the hydraulic gradient, $K(\theta)$ is the water content-dependent hydraulic conductivity; and z and t are depth and time, respectively. The hydraulic potential energy, H, is defined to be the sum of the matric potential energy, h, and the gravitational potential energy. Equation [1] can be solved by either finite-difference or finite-element techniques to provide a model that will predict θ given knowledge of (i) the initial and boundary conditions (at the soil surface, at the bottom of the soil profile) of the system, and (ii) the relationships between K and θ, and θ and h. This latter relationship is also known as the soil–water characteristic curve. As part of the initial condition, the model requires information on the values of $\partial H/\partial z$, which can be estimated either from initial conditions of water content and the soil–water characteristic curve, or from direct measurements of h obtained from either tensiometers or resistance blocks.

Knowledge of the $K(\theta)$ and $h(\theta)$ relationships is therefore crucial in order to use a mechanistic simulation model of soil–water infiltration and redistribution. As will be discussed below, much effort has therefore been invested in estimation of these relationships from soil characteristics and soil properties through pedotransfer functions. Much more effort is required to generalize and extend our experience. However, Eq. [1] can serve as the beginning point for discussion of contemporary models of infiltration and leaching.

The Green-Ampt Infiltration Model

It is important in some applications to partition between infiltration and surface runoff, and to do so in an accurate, yet computationally efficient manner without excess data requirements. Although the Richards equation provides a rather comprehensive method of determining the effects of many interactive factors on infiltration, the hydraulic conductivity function is difficult to measure and is available in the literature for only a few soils. Accordingly, approximate equations for predicting infiltration rates have been proposed, among which the Green and Ampt (1911) model appears to be the most flexible.

The Green and Ampt (1911) model represents a hybrid of mechanistic and nonmechanistic approaches through the assumptions made about the infiltration process. As such, it is operationally simplified from strict application of Eq. [1], but is approximately accurate and useful for land evaluation purposes given adequate input information. As described by Skaggs (1980), it was originally derived from deep homogeneous profiles with a uniform initial water content. Water is assumed to enter the soil as slug flow resulting in a sharply defined wetting front that separates a zone that has been wetted from a totally uninfiltrated zone (Fig. 10-1). Direct application of Darcy's law yields

$$f = K_s \frac{H_2 - H_1}{L_f}, \qquad [2]$$

where f is the infiltration rate (cm/h); L_f is the length of the wetted zone; K_s is the hydraulic conductivity of the wetted or transmission zone; H_1 is the hydraulic head at the soil surface; and H_2 is the hydraulic head at the wetting front.

With several assumptions about the nature of the infiltration process, Eq. [2] can be manipulated to give the Green-Ampt equation

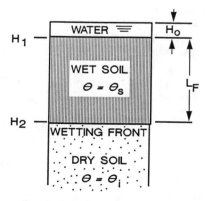

Fig. 10-1. Physical system described by the Green-Ampt (1911) model.

$$f = K_s + K_s M S_{av}/F \quad [3]$$

where S_{av} is the effective suction (matric potential) at the wetting front; M is the initial soil-water deficit (or fillable porosity); and F is the cumulative infiltration from the beginning of the event. For a given soil with a given initial water content, Eq. [3] may be written as

$$f = A/F + B \quad [4]$$

where A and B are parameters that depend on the soil properties, initial water content distribution, and surface conditions such as cover and crusting.

Main and Larson (1973) used the Green-Ampt equation to predict infiltration from steady rainfall. Their results were in good agreement with rates obtained from solutions to Eq. [1] for a wide variety of soil types and application rates. Mein and Larson's results imply that, for uniform deep soils with constant initial water contents, the infiltration rate may be expressed in terms of cumulative infiltration alone, regardless of the application rate. Reeves and Miller (1975) extended this assumption to the case of erratic rainfall where the unsteady application rate dropped below infiltration capacity for a period of time followed by a high intensity application. Their investigations showed that the infiltration capacity could be approximated as a simple function of cumulative infiltration regardless of the application rate vs. time history. These results are extremely important for modeling efforts that seek to separate infiltration and runoff. If the infiltration relationship is independent of application rate, the only input parameters required are those pertaining to the necessary range of initial conditions. On the other hand, a set of parameters covering the possible range in application rates would be required for each initial condition if the infiltration relationship depends on application rate. Clearly, the utility of such a straightforward approach indicates that additional efforts to define A and B from soil survey data would be productive endeavors.

Mechanistic Models of Water Redistribution and Chemical Leaching

The Richards equation provides the beginning point for mechanistic models of water infiltration and redistribution. Recent advances in computing technology make its use possible for multiple year executions without prohibitive requirements for computing time. Information is required, as outlined above, on gradients and relationships between K-θ-h.

Once the solution to Eq. [1] has been used to estimate water content changes with depth and time, it is often necessary to estimate the resulting chemical movement. This is usually accomplished in mechanistic leaching models through use of the convection-dispersion equation, written below to include sorption and transformation processes

$$\frac{\partial(\theta c)}{\partial t} + \frac{\partial(\rho s)}{\partial t} = \frac{\partial}{\partial t}[\theta D(\theta, q)\frac{\partial c}{\partial z} + qc] - \theta kc \quad [5]$$

where new variables c and s are concentrations in the solution and sorbed phases respectively; $D(\theta, q)$ is the diffusion-dispersion coefficient (dependent on θ and water flux, q); k is a first-order decay constant; and ρ is soil bulk density. Estimated values of q (and therefore D) derive from $\theta(z,t)$ obtained from the water-flow model. The solution of Eq. [5], again by numerical methods, depends upon knowledge of θ and q to estimate $c(z, t)$. Without accurate estimation of θ and q, there is little possibility of accurately estimating $c(z, t)$, and so there has been much emphasis placed on directly measuring $K(\theta)$ or deriving it from soil characteristics or properties. Additionally, $c(z, t)$ depends also upon the mediating processes of sorption and transformation, with values of s and k usually obtained from independent studies. However, some attention has also been paid to estimating sorbed phase concentrations from soil characteristics; but quantitative relationships are lacking, little successful attention has been given to relating k to such soil attributes.

Nonmechanistic Models of Water and Chemical Leaching

Mechanistic-leaching models are all similar in concept since there is a general consensus among modelers of unsaturated zone processes that both Eq. [1] and [5] are the pragmatic, acceptable operational approach. The less-mechanistic models, sometimes termed functional models, are more diverse because they embrace not only a variety of methods of simplifying the mechanistic approach, but also a number of alternative methods of simulation.

Functional models of water and chemical movement used for land evaluation generally divide the soil into horizontal layers, and are based upon mass balance principles. These models characterize the water-holding capacity of each layer, usually in terms of θ_{fc}, a field-capacity volume water content. They also require an initial water content condition, θ_i. The difference between θ_i and θ_{fc}, and the depth of water applied, is used to estimate the depth of penetration of the applied water. The deepest possible movement of the chemical, or alternatively the position of maximum chemical concentration, is next estimated. These models are usually recognized to provide less accurate descriptions of water content or chemical concentration distributions due to the greatly simplified assumptions about water and chemical movement. However, they are useful approaches to give approximate estimates for land evaluation questions that require relative, rather than qualitative, comparisons between different scenarios.

Functional models have only modest data demands. All need rainfall/irrigation data and, where appropriate, evaporation data. They also need θ_i, θ_{fc}, and soil porosity or a measurement of the water-saturated soil volume. Variations of layer-type models (such as by Addiscott, 1977) have been developed to describe flow in mobile and immobile soil porosity, and in such cases these models also require knowledge of the soil–water characteristic curve.

ISSUES OF SAMPLING AND DIRECT MEASUREMENT

Many questions related to land evaluation or environmental protection have to be answered using limited resources at such a scale that intensive and systematic sampling to determine the inputs for simulation models is considered to be impossible. Yet, almost every soil scientist is well aware of the inherent spatial variability of land properties and qualities, and can cite example studies that demonstrate that dozens, hundreds, or more samples are needed to reliably estimate $K(\theta)$ and perhaps also $h(\theta)$. This situation is made worse by the nature of most measurement practices beyond those used in standard soil survey exercises. For example, although a wide variety of direct measurement technologies exist for quantifying $K(\theta)$, it appears that these methods are most applicable in either small-scale research studies where great care can be taken by highly trained individuals, or in the occasional (and all too infrequent) case where both trained personnel and substantial resources are available for large-scale field projects.

Direct measurement of $h(\theta)$ or $K(\theta)$ is a relatively straightforward exercise, but often impossible to accomplish at more than a few sites with a limited budget. Selection of an appropriate technique for a limited number of samples requires that morphology and measurement method be strongly linked. These issues have been reviewed in the context of using soil morphological characteristics as a guide to selection of appropriate measurement methods (e.g., Bouma, 1983, 1989), particularly for K_{sat} and the $K(\theta)$ relationship. Classically, when defining physical methods for measuring K, the technical and theoretical aspects are usually emphasized, while considerably less attention is paid to operational considerations of the optimal sample volume, the measurement value obtained, and the resulting estimated spatial variation of K. It is important when measuring such dynamic properties as K_{sat} and $K(\theta)$ that depend greatly on pore geometry to define optimal sample volumes, particularly in soils with natural aggregates (peds) and continuous large pores, such as cracks and root and worm channels.

Several examples illustrate this interdependence for K_{sat} and $K(\theta)$. Discussions on spatial variability of soil hydraulic characteristics have generally focused on mathematical manipulation of measured data, which are usually accepted as presented. However, part of the observed variability may result from using an incorrect method, or an unrepresentative sample volume. Anderson and Bouma (1973) showed that K_{sat} of a silt loam B2t horizon (Typic Argiudoll), as measured in undisturbed soil cores, varied considerably as a function of core height. Average values ranged between 650 cm/d for 5-cm-high cores to 100 cm/d for 17-cm-high cores. The latter had a volume of 270 cm^3 and contained 20 to 30 blocky peds. Similarly, the variance of the population of observed values decreased as sample height increased. The differences were explained by considering the vertical continuity of planar void flow paths (cracks) between peds. As sample height increased, the opportunity increased for constriction or termination of such paths, thereby decreasing the rate of water flow through the sample, as measured by K_{sat}.

Table 1. Dimensions, volumes, and K_{sat} values measured by in situ methods (Lauren et al., 1988).

Column	Dimensions	Volume	Mean	Mode	Median	SD	CV	Number of samples
			——— (cm/d) ———				%	
A	160 by 75 by 20	240 000	21.3	10.3	16.6	16.9	79	37
B	120 by 75 by 20	120 000	13.7	6.4	10.7	11.0	81	36
C	50 by 50 by 20	50 000	14.4	6.3	10.9	12.5	96	37
D	20(diam) by 20	6 283	36.6	6.3	20.3	54.9	150	37
E	7(diam) by 6	884	34.5	4.8	16.3	64.0	186	35

Similar results were reported by Lauren et al. (1988) when measuring steady-state, ponded infiltration rates (as indices of K_{sat}) into the B2t horizon of a Hudson silty clay loam (Glossaquic Hapludalf). Using in situ methods, in which a block of soil was excavated on four vertical sides, left attached to the native soil at the bottom and then encased in gypsum to prevent leakage, the effect of the volume of the block was measured (Table 10-1). The largest of the five volumes (240 000 cm^3) proved operationally too large for reliable measurement of K_{sat}. Two intermediate sizes of 120 000 and 50 000 cm^3 produced measured populations with similar mean, mode, median and standard deviation. However two additional smaller sizes (6283 and 884 cm^3) gave much larger mean values and larger standard deviations. The authors interpreted this as evidence that sample volume must be greater than a minimum number of peds in order to obtain representative measurements of K_{sat} that were useful in characterizing soil variability. This evidence was consistent with results of Bouma et al. (1979), who found when studying K_{sat} in heavy clay soils with large prisms that a minimum volume of 16 to 20 L were needed to measure K_{sat} with the column method. Bouma (1983) interprets this further to indicate that a minimum of 20 peds must be encompassed in order for any one sample to be a representative sample useful for measuring K_{sat}.

For purposes of simulation modeling, it is often the objective to measure not only K_{sat}, but also to estimate the balance of the $K(\theta)$ relationship. This can be accomplished by either field or laboratory techniques (Klute, 1986), but due to the relative sophistication of both theoretical and practical aspects of the field methods and the commercial availability of specially designed apparatus, laboratory determination is most often used. While accurate measurements of K_{sat} are very dependent on the issues outlined above, unsaturated K values are less sensitive to sample volume. Since macroporosity, whether it be wormholes, planar voids, or root channels, drains readily with small increments of applied pressure, there is little effect of these spatially distributed, variable geometries upon hydraulic conductivities at most water contents less than saturation. It is still important to work with undisturbed samples that preserve natural pore geometries, but it is less important to use sample sizes of 20 peds or more (Bouma, 1983). Of greater consequence is the accurate estimation of $K(\theta)$ at several water contents between saturation and θ_{fc} (approximately 20 kPa), for it is in this region that a large portion

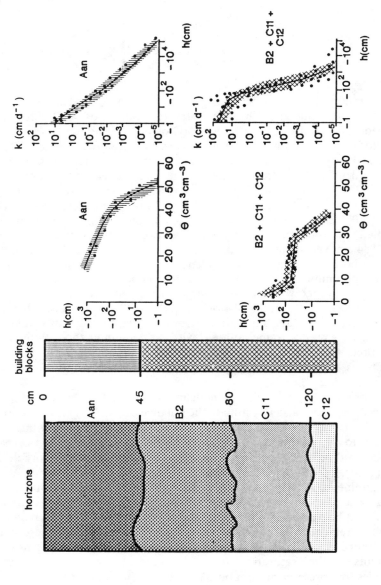

Fig. 10–2. Translation of mapped soil horizons into corresponding soil physical properties.

of water redistribution and resulting chemical leaching occurs. Inaccuracy in this region of the $K(\theta)$ curve, even with accurate estimation of K_{sat}, can greatly impact the accuracy of simulated water and chemical fluxes.

Direct measurement of the soil–water characteristic curve is possible under either field or laboratory conditions, often with less investment of labor and resources than needed to measure $K(\theta)$. Many of the same issues that influence choice of a $K(\theta)$ measurement method influence selection of an $h(\theta)$ method as well, although it takes only a tensiometer and a simultaneous measure of θ (e.g., neutron probe, gravimetric sampling) to measure $h(\theta)$ in situ to about -70 kPa. The relative ease of estimating $h(\theta)$ has led to the development of several theoretical models that presume measurement of K_{sat} and $h(\theta)$, followed by fitting of a function to the $h(\theta)$ curve. Parameters from this function can then be used with K_{sat} to calculate the $K(\theta)$ curve. These methods appear to be quite useful in many cases for developing input to simulation models, although when the value of K_{sat} is greatly influenced by the presence of macroporosity, there does not yet appear to be a fully tested and proven method of estimating $K(\theta)$ for $\theta_{sat} < \theta < \theta_{fc}$. More studies are needed on this issue.

There appears to be great potential in use of $h(\theta)$ as a means of summarizing soil characteristics in a way that relates dynamic soil qualities to static morphological categories. For example, Wosten et al. (1985) characterized the physical soil properties of 12 different pedogenic soil horizons, in an area of 800 ha, to find that only six of those were significantly different from a physical point of view. This illustrates the important phenomenon that pedogenic differences do not necessarily correspond with functional differences (Fig. 10-2). It also demonstrated that by using pedogenic soil horizons as carriers of information, the basic problem of extrapolation of point to area data can be solved. This approach is particularly valuable when used in conjunction with the development of pedotransfer functions, as discussed below.

ESTIMATING INPUT DATA FOR SIMULATION MODELING

It is often not possible or practical to directly measure the soil characteristics and properties required to execute a simulation model. Particularly soil data, such as $K(\theta)$ or $h(\theta)$ relationships, are often lacking. Although the modeler can use experience or intuition to define these relationships, more quantitative and objective methods are preferred. These methods, once they are established as reliable and accurate approaches, provide the opportunity to estimate key relationships that are difficult, expensive or time consuming to measure. If such methods can be developed, they greatly simplify the minimum taxonomic data required by modelers.

One approach to development of such methods recognizes that soil survey activities have resulted in the measurement and tabulation of a wide range of soil characteristics including texture, structure, organic matter content, and cation exchange capacity. Although these data alone are insufficient to

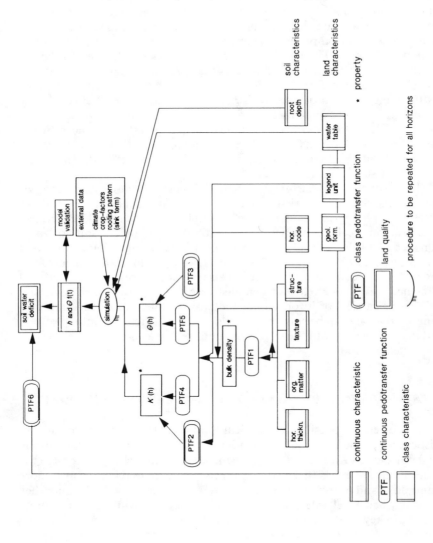

Fig. 10-3. Flow diagram for the land quality Soil Water Deficit (Bouma & van Lanen, 1987).

meet the needs of simulation modelers, a major challenge is to "translate" them into the data that modelers require. This is a particularly relevant issue for $h(\theta)$ and $K(\theta)$ relationships, which will probably never be directly measured in the large number of field locations at which modelers would prefer to have information.

Procedures to obtain these "translations" have been discussed and extended into practical use by Bouma and van Lanen (1987). The term "pedotransfer functions" (PTF) has been defined as an expression that relates different soil characteristics and properties with one another or to land qualities. These functions are quite useful if it is clearly recognized that they apply only within the limits of the data used in their development. Two types of pedotransfer functions are distinguished: those that are continuous (e.g., using percentage clay, percentage organic matter, etc.) and those that relate to distinct classes (soil types, horizon designations). Several examples will illustrate the usefulness of such functions.

Consider first the estimation of the soil-water deficit, a land quality important in crop production. Simulation of water regimes and crop response requires $K(\theta)$ and $h(\theta)$ data, water table levels, rooting depth and climatic data. Figure 10-3 (from Bouma, 1989) indicates that six soil characteristics, each contained in soil survey reports, can be used as starting points to estimate the soil-water deficit. Four of these are continuous characteristics and two are class characteristics. Additionally the land characteristic on groundwater class defines the mean highest and lowest level of the water table.

Simulation of the soil-water deficit using a mechanistic model based on Eq. [1] requires use of $h(\theta)$ and $K(\theta)$ to estimate $\theta(z, t)$ in the presence of crop extraction during a growing season. As an example of the use of pedotransfer functions, input to the model on the $h(\theta)$ and $K(\theta)$ relationship can be developed using pedotransfer function PTF1 that has been derived by regression analysis to relate soil bulk density to organic matter and texture. Then, the bulk density is used further to estimate $K(\theta)$ and $h(\theta)$ through transfer functions PTF4 and PTF5. These functions can be of the form $\theta(h = -100 \text{ cm}) = b_0 + b_1 L + b_3 S_y$ where b_i are coefficients determined by regression analysis of numerous samples; L is percentage clay and S_y is 1/bulk density. In practical terms, PTF1 will probably seldom be used, as bulk density is easy to measure; and hence PTF4 and PTF5 can be used entirely with measured bulk density values. Other procedures, including estimation of the entire shape of the $K(\theta)$ or $h(\theta)$ curves, rather than points on the curve, have also been developed (Ghosh, 1980; Clapp & Hornberger, 1978; Rawls & Brakensiek, 1982). For example, Cosby et al. (1984), through regression analysis of nearly 1500 soil samples presented by Holtan et al. (1968) and Rawls et al. (1976), estimated such curves and the variability associated with them. They first assumed that

$$h = a(\theta/\theta_s)^{-b} \quad [6]$$

and

$$K = K_{\text{sat}} (\theta/\theta_s)^{2b+3} \quad [7]$$

where a and b are parameters describing the shape of the soil–water characteristic curve.

Their regression analysis gave values for the mean (μ) and standard deviation (σ) of each variable in Eq. [6] and [7] as

$$\mu_{\log a} = -0.0095(\% \text{sand}) - 0.0063(\% \text{silt}) + 1.54; \quad r^2 = 0.850 \quad [8a]$$

$$\mu_b = 0.1570(\% \text{clay}) - 0.0030(\% \text{sand}) + 3.10; \quad r^2 = 0.966 \quad [8b]$$

$$\mu_{\theta_s} = -0.142(\% \text{sand}) - 0.037(\% \text{clay}) + 50.50; \quad r^2 = 0.785 \quad [8c]$$

$$\mu_{\log K_{sat}} = 0.0126(\% \text{sand}) - 0.0064(\% \text{clay}) - 0.60; \quad r^2 = 0.872 \quad [8d]$$

$$\sigma_{\log a} = -0.0026(\% \text{silt}) + 0.0012(\% \text{clay}) + 0.72; \quad r^2 = 0.111 \quad [8e]$$

$$\sigma_b = 0.0492(\% \text{clay}) + 0.0144(\% \text{silt}) + 0.92; \quad r^2 = 0.584 \quad [8f]$$

$$\sigma_{\theta_s} = -0.0805(\% \text{clay}) - 0.0070(\% \text{sand}) + 8.23; \quad r^2 = 0.574 \quad [8g]$$

$$\sigma_{\log K_{sat}} = 0.0032(\% \text{silt}) + 0.0011(\% \text{clay}) + 0.43; \quad r^2 = 0.403. \quad [8h]$$

Equations (6–8) can be used to estimate both $h(\theta)$ and $K(\theta)$ given knowledge only of texture and organic matter. Additionally, with proper techniques (Addiscott & Wagenet, 1985), values of a, b, θ_s and K_{sat} can be repeatedly selected from their respective frequency distributions (given by Eq. [8]) to construct possible $h(\theta)$ or $K(\theta)$ relationships in an area in which such relationships are known to be spatially variable, but where only soil survey information is available. This would form the beginning point for simulation of water and chemical fluxes in spatially variable soil, using only information developed in the process of classification. Further refinement of such methods, including analyses of other soils and better estimation of $h(\theta)$ and $K(\theta)$ in the wet region (Cosby et al. [1984] did not consider $h > -20$ kPa) is needed.

An alternative approach to estimating $K(\theta)$ and $h(\theta)$ is represented in Fig. 10-3 using the class pedotransfer functions PTF2 and PTF3. By basing the estimation process upon mapped soil horizons, and if the map unit delineations are homogeneous, the problem of extrapolation of point data to area data is solved, since horizons have a geographic distribution defined by mapped soil series. Additionally, as shown by Wosten et al. (1985), pedogenic differences do not necessarily correspond with functional differences for $h(\theta)$. Similar results have been presented for K_{sat} (McKeague et al., 1984) and phosphate sorption capacity (Breeuwsma, 1986). The key result in all cases is that cumbersome direct measurement of K_{sat}, $K(\theta)$, $h(\theta)$ and phosphate sorption could be replaced by a procedure using available data assembled in the context of pedology studies. Many other functional characterizations can almost certainly be derived, for example, nitrification and denitrification potential based on organic matter content, mechanical stability based on texture and structure and pesticide mobility based partially on clay and organic matter content. The opportunities in this area are relatively unexplored.

A final option for use of taxonomic information for simulation modeling purposes is represented by PTF6 in Fig. 10-3. It is possible to estimate the soil–water deficit by direct consideration of the soil series and water table fluctuations class. The PTF6 assumes that several calculations have been made for a particular soil series, allowing predictions for an identical series elsewhere. This must be cautiously accomplished, as subsoil textural variation within a given soil series is possible. However, where applicable, PTF6 allows rapid development of derived maps (e.g., soil–water deficit) from soil maps, through the use of existing soil series and simulation modeling of relevant example cases.

SOIL MANAGEMENT AND SIMULATION MODELING

The influence of soil management upon the process of simulation modeling for land evaluation has been approached from different directions by U.S. and European scientists. In the USA, a soil phase is defined through the application of a range of nontaxonomic criteria such as slope, degree of erosion, and near-surface texture that are considered to have management implications. Changes in soil properties with time or management are not directly considered in defining the soil phase. Yet models are applied to individual soil phases to estimate the influence of the nontaxonomic criteria on water and chemical movement. Modelers also consider that management practices can influence soil properties, such as structure near the surface. Improvements in the models to consider such issues are currently the focus of several research efforts.

The beginning point for definition of the soil phase in the Netherlands (and perhaps elsewhere in Europe) has been the recognition that the key relationships used in simulation modeling, $K(\theta)$ and $h(\theta)$, are very much a function of the management to which the soil has been subjected. It is a basic assumption that changes in $K(\theta)$ and $h(\theta)$ will result from changes in soil structure due to tillage, compaction or incorporation of organic residues. This has effectively been treated by defining "soil structure phases of well-defined soil series." A good example is presented in Fig. 10-4, which presents two different phases of a Dutch soil (Typic Haplaquent). The soil

Fig. 10-4. Vertical sections through two Dutch soils with identical medium-textures but different soil structures resulting from different soil and plant management. (Scale unit = 10 cm); photograph provided by Department of Soil Morphology, Netherlands Soil Survey Institute.

on the left has well-rounded natural aggregates ("peds") due to high biological activity under grassland vegetation—it contains many roots. The soil on the right has been compacted by machinery on arable land, forming large angular peds that plant roots cannot penetrate. Soil phases in this case distinguish identical soils from a pedogenetic point of view that may behave differently in terms of physical, chemical or biological processes due to the effects of management practices upon soil structure. As with the U.S. approach, it will become ever more important to distinguish soil phases as simulation models are used for land evaluation purposes and hydraulic properties are estimated from soil survey data bases. The conditions under which the data base information was developed could greatly alter the usefulness of it for estimating $K(\theta)$ or $h(\theta)$, depending upon the management practices, or the management criteria used in the classification process to distinguish soil phases.

This aspect is important, because soil morphological studies in the past have often suffered from a "classification syndrome," yielding highly detailed descriptions that focus more on morphological characterization than on functional characterization. It appears that in this respect, much of the classifiers' effort and focus in the past has been directed inward to a peer audience of other classifiers, rather than outward to a larger group of users, who often look at strictly morphological characterizations and rightly ask "How do I use this information to predict soil behavior and performance?"

The need for a linkage between morphology and function in the presence of soil phases was clearly illustrated by Kooistra et al., (1984, 1985) and van Lanen et al. (1987). The former defined a limited number of structure types for a Dutch sandy loam soil (Typic Haplaquent). These types had significantly different $K(\theta)$ and $h(\theta)$ curves, and were so different from a soil morphological point of view that different observers could arrive at identical classifications. However, when simulation modeling was used as a tool for determining such land qualities as moisture availability, aeration status and trafficability, considerable differences were found between grassland and tilled soil. Therefore, an important next step in using taxonomic data for simulation modeling is to inseparably link soil structure and hydraulic properties, and to further distinguish soil phases as manifestations of the effects of soil management practices. Proper consideration of soil phases as a factor in estimation of hydraulic properties from soil survey data is another of the topics on which further research is needed.

EXAMPLE SIMULATION MODELS AND THEIR MINIMUM DATASET

The minimum data set required for simulation modeling varies according to the model used. Three examples of models useful for different purposes, but all of which utilize information either found in or derived with pedotransfer functions from soil survey data bases will illustrate the process.

The USDA Water Erosion Prediction Project model (WEPP), intended for application to hillside erosion is presented by Lane and Nearing (1989). The erosion process is divided into interrill and rill erodibility. The interrill process involves delivery of sediment to rills and this delivery is considered proportional to the square of rainfall intensity. Rill erodibility pertains to soil detachment by concentrated flow, and is commonly defined as the increase in soil detachment per unit increase in shear stress as applied by clear water. For there to be net soil detachment the hydraulic shear stress must exceed the shear stress at which detachment initiates, referred to as the critical shear stress.

Input to the model is based upon measurements of both interrill and rill erodibility, which have been made on freshly tilled cropland soil and on rangeland from which vegetation had been removed but with little soil disturbance. The measured values are properties to predict the erodible quantities. These quantities are adjusted for vegetal cover, water consolidation, and other processes that pertain to the near surface.

The model is dependent on runoff estimates that are generated from infiltration estimates provided by the Green-Ampt infiltration model (Eq. [3]). The quantities required for Green-Ampt are available porosity, wetting front matric potential, and saturated hydraulic conductivity. Available porosity is the difference between the total porosity and the water volume present with a correction for pore space that would not be filled with water at zero suction. The water volume present is provided from the balance between precipitation and evapotranspiration. Near-surface changes in bulk density and in total porosity are modeled based on the kind of tillage operation and rainfall since the last tillage operation. The wetting-front matric potential is obtained from soil textural components and total porosity. Saturated hydraulic conductivity is predicted from the affected porosity, bulk density, residual water content, and particle-size components. Changes in the near-surface bulk density related to use and rainfall affect the saturated hydraulic conductivity. Adjustments for frozen soil can be made. Protocols are given for adjustments related to the crust and to near-surface macroporosity.

The DRAINMOD is a computer simulation model for design and evaluation of water management for soils with high water tables (Skaggs, 1980). The soil-related inputs are considered here. The Green-Ampt equation for prediction of infiltration is again employed. The parameters are commonly estimated from standard soil characterization data, but they may be evaluated from experimental infiltration measurements. When the rainfall rates exceed the infiltration rate calculated by the Green-Ampt equation, the excess water is assigned to the ground surface. When the ground-surface storage is exceeded runoff occurs. For most practical drainage situations it is advantageous to maximize the runoff and reduce the depressional storage on the ground surface. Depressional storage is modeled based on land-use treatment information. Subsurface drainage through tiles or ditches depends on the saturated hydraulic conductivity, and the spacing and depth of the soil alterations designed to change the drainage. Required inputs are the midpoint water table height above the drain, the effective lateral saturated hydraulic conductivity, the distance between drains, and the equivalent direction for radial flow near drains.

DRAINMOD also requires saturated hydraulic conductivity, the soil-water characteristic curve, and the relationship between water table movement and the volume of water removed or added. The slope of the plot of drainage volume vs. water table depth is the drainable porosity that can be calculated from a characteristic curve. Upward flow of water to the bottom of the root zone is considered through several approaches that range in sophistication. For example, in one approach a critical limiting depth is assigned below which water will not move upward to the root zone. Water is then assumed to move upward through the root zone at the potential evapotransportation rate. Texture and saturated hydraulic conductivity estimates are employed to define the critical depth. Finally, trafficability is also considered from the point of view of the timing of tillage and other farming operations.

The LEACHM (Leaching Estimation and CHemistry Model) is a one-dimensional model of water infiltration and redistribution and chemical transport within the unsaturated zone (Wagenet & Hutson, 1989). It is a general acronym that refers to three versions of a simulation model that utilize similar numerical solution schemes to simulate water flow and (i) N transport and transformation, (ii) pesticide displacement and degradation, and (iii) transient movement of inorganic ions (Ca, Mg, Na, K, SO_4, Cl, CO_3, HCO_3). These models are intended to be applied to field situations.

Estimates of plant growth and absorption of water and solutes by plant roots are included in all models, together with a flexible means of describing rainfall and surface evaporation of water. A heat flow simulation producing soil temperature profiles is also included.

The finite-difference numerical procedures allow description of a wide range of field conditions, flexibility in simulating layered or nonhomogeneous profiles, and orderly and self-explanatory input and output tables. The models are organized on a modular basis, with a main program that initializes variables, calls subroutines and performs mass balancing. Subroutines deal with data input and output, time-step calculation, evapotranspiration, water flow, solute movement, sources, sinks (degradation, volatilization) and chemistry, leaf and root growth, temperature, and solute absorption by plants. Segregation of each of these processes into subroutines called by the main program enables any subroutine to be replaced by an improved or different formulation if desired. The modeling approach, as with all numerical methods, divides the profile into a number of horizontal segments of equal thickness (Fig. 10–5). Simulations begin by defining a set of initial conditions (outlined below). The soil need not be homogeneous in the vertical direction. Plants can be present or absent. If present, crop cover and root expansion can be simulated, or a static, established root system and crop cover can be defined. The following inputs are required:

1. Soil properties and initial conditions for each soil segment: (i) water content or matric potential, (ii) hydrological constants for calculating retentivity and hydraulic conductivity or, alternatively, knowledge of the particle-size distribution and an appropriate pedotransfer function, (iii) chemical contents and soil chemical properties appropriate for the solute being considered.
2. Soil surface boundary conditions of: (i) irrigation and rainfall amounts and rate of application, (ii) mean temperature and diurnal amplitude for each period regarded as having a constant temperature regime (only if a temperature simulation is required), (iii) pan evaporation (weekly totals).
3. Crop details (if it is assumed that no crops are present, a control variable allows bypass of the plant-related subroutines): (i) time of planting, (ii) root and crop maturity and harvest, (iii) root and cover growth parameters, (iv) a pan factor for adjusting pan evaporation to potential crop evapotranspiration, and (v) lower soil and plant water potentials for water extraction by plants.
4. Other constants used in determining bottom boundary conditions, time steps, diffusion coefficients and output details. Some of these constants rarely require alteration, but are listed in the data files to define their value for the user and provide the opportunity for change.

Output of the model at specified print intervals or times includes:

1. Hydraulic conductivities and water contents for each layer of the soil at selected soil–water matric potential values of 0, -3, -10, -30, -100, and -1500 kPa.

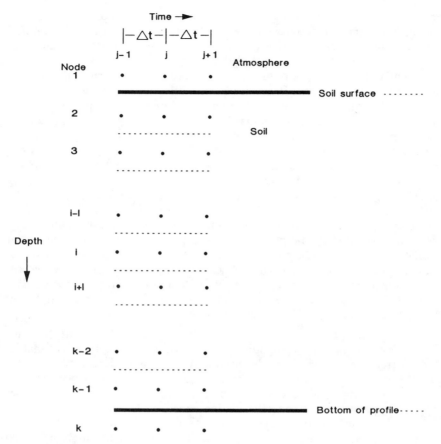

Fig. 10-5. Definition of nodes in LEACHM. Node spacing, usually uniform, can be designed to correspond to mapped horizonation. This allows soil properties and input to simulation models to change with depth.

2. Cumulative totals and mass balances of water and all solutes considered in the model being used. This includes the amount of material initially in the soil profile, currently in the profile, the simulated change, additions, losses, and a composite mass error.
3. A summary by depth of water content, matric potential, water flux between layers, soil temperature, evapotranspiration, and mass and concentration of individual chemical species.
4. A summary by depth of root density, water uptake, and solute uptake.

CONCLUSIONS

1. Soil surveys form a unique, easily accessible source of soil data that can be used to obtain basic information needed for simulation models of

water and nutrient movement and associated crop growth. Use of these models is necessary for modern land evaluation.

2. A concentrated effort should be made to further develop pedotransfer functions that relate available soil data to information needed for modeling. Both continuous and class pedotransfer functions need to be developed, as discussed in this chapter. Development of pedotransfer functions is crucial because measurement of model parameters for every site to be characterized is prohibitive in terms of cost.

3. Soil survey information can also be used to improve physical and chemical sampling and monitoring techniques by considering representative elementary volumes of soil samples and by placement of physical monitoring equipment based on an analysis of morphological soil features, such as occurrence of peds, macropores and specific soil horizons.

4. Methods of simulation modeling must be developed that consider changes in soil properties following different types of soil management within well-defined soil series. Similarly, soil properties and land characteristics must be defined with reference to both the influence of past management upon the property, and the effects of nontaxonomic variables on soil management.

5. There is a need to develop better physical models of field situations. Necessary improvements include better prediction of water infiltration by Green-Ampt techniques and development of methods useful in cases where Green-Ampt is inappropriate (e.g., macropores). Additionally, a scheme is needed to predict the resistance of the near-surface to wind and water erosion. Current techniques lack sophistication and information about changes with use and the effect of time within a use on the physical organization of the near-surface and the related resistance to erosion.

REFERENCES

Addiscott, T.M. 1977. A simple computer model for leaching in structured soils. J. Soil Sci. 28:554–563.

Addiscott, T.M., and R.J. Wagenet. 1985. A simple method for combining soil properties that show variability. Soil Sci. Soc. Am. J. 49:1365–1369.

Anderson, D.L., and J. Bouma. 1973. Relationship between hydraulic conductivity and morphometric data of an argillic horizon. Soil Sci. Soc. Am. Proc. 37:408–413.

Bouma, J. 1983. Use of soil survey data to select measurement techniques for hydraulic conductivity. Agric. Water Manage. 6:177–190.

Bouma, J. 1984. Using soil morphology to develop measurement methods and simulation techniques for water movement in heavy clay soils. p. 298–316. *In* J. Bouma and P.A.C. Raats (ed.) Proc. of ISSS Symp. Water and solute movement in heavy clay soils, Wageningen, The Netherlands. 27–31 Aug. 1984. Publ. 37. Inst. for Land Reclamation and Improvement, Wageningen, The Netherlands.

Bouma, J. 1989. Using soil survey data for quantitative land evaluation. p. 177–213. *In* B.A. Stewart (ed.) Advances in soil science, Volume 9, Springer-Verlag, New York.

Bouma, J., L.W. Dekker, and J.C.F.M. Haans. 1979. Drainability of some Dutch clay soils: A case study of soil survey interpretations. Geoderma 22:193–203.

Bouma, J., and H.A.J. van Lanen. 1987. Transfer functions and threshold values: From soil characteristics to land qualities. p. 106–110. *In* K.J. Beek et al. (ed.) Proc. ISSS/SSSA Workshop on Quantified Land Evaluation, Washington, DC. 27 Apr.–2 May 1986. Int. Inst. for Aerospace Surv. and Earth Sci. Publ. no. 6. ITC Publ. Enschede, The Netherlands.

Breeuwsma, A., J.H.M. Wosten, J.J. Vleeshouwer, A.M. van Slobbe, and J. Bouma. 1986. Derivation of land qualities to assess environmental problems from soil surveys. Soil Sci. Soc. Am. J. 50:186-190.

Clapp, R.B., and G.M. Hornberger. 1978. Empirical equations for some soil hydraulic properties. Water Resour. Res. 14:601-604.

Cosby, B.J., G.M. Hornberger, R.B. Clapp, and T.R. Ginn. 1984. A statistical exploration of the relationships of soil moisture characteristics to the physical properties of soils. Water Resour. Res. 20:682-690.

Food and Agriculture Organization. 1976. A framework for land evaluation. Soils Bull. 32. FAO. Rome.

Ghosh, R.F. 1980. Estimation of soil moisture characteristics from mechanical properties of soils. Soil Sci. 130:60-83.

Green, W.H., and G.A. Ampt. 1911. Studies on soil physics: I. Flow of air and water through soils. J. Agric. Sci. 4:1-24.

Holtan, H.N., C.B. England, G.P. Lawless, and G.A. Schumaker. 1968. Moisture-tension data for selected soils on experimental watersheds. Rep. ARS 41-144. ARS, Beltsville, MD.

Klute, A. (ed.) 1986. Methods of soil analysis. Part 1. Agronomy Monogr. 9. ASA, SSSA, Madison, WI.

Kooistra, M.J., J. Bouma, O.H. Boersma, and A. Jager. 1984. Physical and morphological characterization of undisturbed and distrubed plough pans in a sandy loam soil. Soil Tillage Res. 4:405-417.

Kooistra, M.J., J. Bouma, O.H. Boersma, and A. Jager. 1985. Soil structure variation and associated physical properties of some Dutch Typic Haplaquents with sandy loam texture. Geoderma 36:215-229.

Lane, L.J., and M.A. Nearing. 1989. USDA-Water erosion prediction project: Hillslope profile model documentation. NSERL Rep. no. 2. USDA-ARS Natl. Soil Erosion Res. Lab., West Lafayette, IN.

Lauren, J.G., R.J. Wagenet, J. Bouma, and J.H.M. Wosten. 1988. Variability of saturated hydraulic conductivity in a Glossaquic Hapludalf with macropores. Soil Sci. 145:20-28.

McKeague, J.A., R.G. Eilers, A.J. Thomasson, M.J. Reeve, J. Bouma, R.B. Grossman, J.C. Favrot, M. Renger, and O. Strebel. 1984. Tentative assessment of soil survey approaches to the characterization and interpretation of air-water properties of soils. Geoderma 34:69-100.

Mein, R.G., and C.L. Larson. 1973. Modeling infiltration during a steady rain. Water Resour. Res. 9:384-394.

Rawls, W.J., and D.L. Brakensiek. 1982. Estimating soil water retention from soil properties. J. Irrig. Drain. Div. Am. Soc. Civ. Eng. 108:166-171.

Rawls, W.J., P. Yates, and L. Asmussen. 1976. Calibration of selected infiltration equations for the Georgia Coastal Plain. Rep. USDA-ARS-S-113. ARS, Beltsville, MD.

Reeves, M., and E.E. Miller. 1975. Estimating infiltration for erratic rainfall. Water Resour. Res. 11:102-110.

Skaggs, R.W. 1980. DRAINMOD. Reference report. Methods for design and evaluation of drainage-water management systems for soils with high water tables. USDA-SCS South Natl. Tech. Center, Forth Worth, TX.

van Lanen, H.A.J., M.H. Bannick, and J. Bouma. 1987. Use of simulation to assess the effect of different tillage practices on land qualities of sandy loam soil. Soil Tillage Res. 10:347-361.

Wagenet, R.J., and J.L. Hutson. 1989. LEACHM: Leaching estimation and chemistry model. A process-based model of water and solute movement, transformations, plant uptake and chemical reactions in the unsaturated zone. Version 2.0. Continuum Vol. 2. Water Resour. Inst., Cornell Univ., Ithaca, NY.

Wosten, J.H.M., J. Bouma, and G.H. Stoffelsen. 1985. The use of soil survey data for regional soil water simulation models. Soil Sci. Soc. Am. J. 49:1238-1245.

11 Quantifying Map Unit Composition for Quality Control in Soil Survey

L. C. Nordt, John S. Jacob, and L. P. Wilding
Texas A&M University
College Station, Texas

ABSTRACT

Characterizing soil spatial variability in soil surveys is critical for maintaining user confidence and soil survey credibility. The purpose of this investigation was to examine the use of USDA-SCS map unit concepts for portraying soil spatial variability and then demonstrate methods for displaying map unit composition in soil survey reports. We examined the composition of four consociation map units in Brazos County, located in the Tertiary Gulf Coastal Plain of Texas. The point-intercept transect method and binomial probability and classical statistical procedures were employed. In all but one case, sufficient observations were obtained to estimate the compositions within ±15% of the mean at the 80% probability level. Results showed that taxonomic purities of the reference taxa ranged from highs of 49 and 45% for the Crockett (Udertic Paleustalfs) and Spiller (Udic Paleustalfs) series, respectively, to lows of 21 and 11% for the Robco (Aquic Arenic Paleustalfs) and Rader (Aquic Paleustalfs) series, respectively. The concepts of similar and dissimilar soils, as they related to the major land uses of the area, were used to establish interpretive groupings. Subsequent interpretive purities of the Crockett and Spiller improved to 86 and 81%, while the Robco and Rader improved to 52 and 48%, respectively. The Robco and Rader did not meet the criteria for consociations and were therefore designed as multitaxa map units. Because the named components of these units did not occur in a consistent, coterminous pattern from delineation to delineation, a complex map unit could not be designed. We are therefore proposing the implementation of a new map unit concept to accommodate these conditionalities. In addition, we propose that tables be inserted in future soil surveys that will statistically rate, at given probability levels, the compositional purities of the map units in the survey area.

At the inception of the National Cooperative Soil Survey Program (NCSSP), the USA was predominantly an agricultural society, with mapping and interpretation priorities set accordingly. Land use today is frequently more intensive and, as a result, there is greater demand for more precise statements

---

Copyright © 1991 Soil Science Society of America, 677 S. Segoe Rd., Madison, WI 53711, USA. *Spatial Variabilities of Soils and Landforms.* SSSA Special Publication no. 28.

about soil–landscape distributions and soil behavior. Users want to know how much and what kind of variability may be expected in delineations of soil map units so that management decisions can be made with a higher degree of confidence. For example, we should be able to estimate the mean composition of the reference taxon (taxa) of a map unit at specified probability levels. Furthermore, probability estimates of soil variability and individual soil properties are necessary if we are to extrapolate properties from one delineation to the next (Wilding, 1988).

Characterizing soil spatial variability in soil surveys has been a persistent problem (Miller et al., 1979). Numerous studies have shown that compositional purities for taxonomic map units are commonly less than 50% (Powell & Springer, 1965; Wilding et al., 1965; McCormack & Wilding, 1969; Crosson & Protz, 1974; Amos & Whiteside, 1975; Bascomb & Jarvis, 1976; Ransom et al., 1981; Edmonds & Lentner, 1986; Mokma, 1987). Some of the early studies demonstrated that map unit purities of 85%, required by the *Soil Survey Manual* (Soil Survey Staff, 1951), were rarely attainable. As a result, *Soils Memorandum 66* (Soil Survey Staff, 1967) was instituted that permitted up to 50% inclusions, providing the inclusions were similar in behavior to the reference taxon (taxa) for which the map unit was named. In addition, no more than 25% dissimilar inclusions were permitted, of which only 15% were allowed to be limiting.

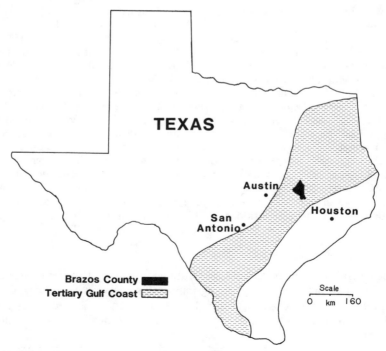

Fig. 11-1. Location of transect area in Brazos County, TX.

The broadness of Memorandum 66 resulted in most map units being designed as single taxon units known as consociations (Soil Survey Staff, 1983). Our experience in mapping soils in the Tertiary Gulf Coastal Plain of Texas (Fig. 11-1) leads us to believe that the designation of consociation in many map units, unintentionally conceals significant variability (and frequently intentionally for mapping convenience). When users encounter this unexplained variability, confidence in the soil survey is questioned. Map-unit compositional studies, in regions of soil complexity similar to that of the Tertiary Coastal Plain, have shown that dissimilar inclusions commonly exceed the allowable 25% and therefore warrant the design of multitaxa map units (McCormack & Wilding, 1969; Amos & Whiteside, 1975; Ransom et al., 1981; Mokma, 1987).

The purpose of this investigation was to quantify and evaluate the composition of four map units in Brazos County, TX (Fig. 11-1), and then to examine how the data could be conveyed to the user, both through map unit design as established by the *National Soils Handbook* (Soil Survey Staff, 1983) and through a compositional rating system that we propose for use in soil survey reports.

MATERIALS AND METHODS

Study Area and Soils

Brazos County is located in the Ustic moisture regime of the Tertiary Coastal Plain of east central Texas (Fig. 11-1) and is underlain by weakly consolidated clays, silts and sands deposited as a series of interfingering fluvial/deltaic stream systems (Bureau of Economic Geology, 1981; Sellards et al., 1932). We have found, through field mapping, that soils formed from these parent materials exhibit a high degree of close-range spatial variability.

Brazos County and most of the Tertiary Gulf Coastal Plain are dominated by pastureland and hayland vegetated mostly by improved varieties of bermuda grass (*Cynodon dactylon*). Soil properties such as surface thickness and texture, and subsoil texture, are important for interpretations of these uses. The increasing spread of urbanization has now created a demand for more precise statements of interpretations for septic tanks, building and road foundations, and recreational facilities. Shrink-swell properties, in addition to the above properties, are most important for these interpretations (Klich et al., 1990).

Four consociations were chosen for compositional evaluation (Table 11-1). These units were chosen because of the large acreage (>55 000 ha) they comprise in Brazos and several surrounding counties. All reference taxa are classified as Paleustalfs (Soil Survey Staff, 1987) at the great group level with taxonomic differences represented at the subgroup level and lower. The

Table 11-1. Reference taxa and their taxonomic classifications.

Crockett	Fine, montmorillonitic, thermic Udertic Paleustalf
Rader	Fine-loamy, mixed, thermic Aquic Paleustalf
Robco	Loamy, mixed, thermic Aquic Arenic Paleustalf
Spiller	Fine, mixed, thermic Udic Paleustalf

Spiller, Robco and Crockett map units are on gently rolling upland landscapes while Rader is on nearly level, fluvial terraces.

Sample Methods and Statistical Analysis

Five delineations from each map unit were randomly selected from completed field sheets (1:20 000). The populations of delineations were randomly stratified by map sheet. Generally, only one delineation was selected from a field sheet; but if it became necessary to select two delineations on a single sheet because of access problems or limited areal distribution, the sampled delineations were arbitrarily separated by a minimum distance of 1.5 km.

The point-intercept transect method was employed (Wilding & Drees, 1983), using a fixed observation interval of approximately 80 m for each transect. This interval was chosen based on spacings needed to achieve independence of observations by using geostatistical analysis on similar soils in the area (Brubaker, 1989). The first and last observation sites were located so that they were not closer than 40 m to the delineation margin. Transects on aerial photographs were drawn normal to the drainage pattern of each delineation.

Following standards and procedures of the *Soil Survey Manual* (Soil Survey Staff, 1981), soil descriptions were written from 4-cm cores taken with a truck-mounted hydraulic probe unit. Each pedon was classified in the field to the family level, and to the series level when possible. Clay percentages of all horizons were estimated by two field soil scientists. When there was some question as to the correct texture, selected pedons or satellite horizons were sampled for particle-size analysis using the hydrometer method of Day (1965). Calibration of field estimates of clay content with laboratory analyses has been followed over the course of the soil survey program. With such calibrations, our field error for estimating clay percentages was within about $\pm 2\%$ of the laboratory-determined values.

Binomial probability and classical statistical procedures were employed to estimate the map unit population parameters (Arnold, 1977, 1981). To estimate composition of the four reference taxa, the means, standard deviations and confidence limits (80% probability level) were calculated. The same parameters were then calculated for the reference taxon plus similar soils (interpretive composition). Finally, the number of observations needed to estimate these compositions within 15% of the mean at the 80% probability level was also calculated.

Table 11-2. Properties considered interpretatively dissimilar from respective reference taxa categorized by taxonomic level.

Taxonomic level	Dissimilar property
Order	None
Suborder	Presence of aquic moisture regime
Great Group	Presence or absence of abrupt textural boundary between surface and subsoil horizons (E/Bt contacts)
Subgroup	Surface thickness outside taxonomic range by ±12 cm, and presence or absence of vertic and nonvertic properties
Family	Clay content outside taxonomic range by ±2%
Series	Surface thickness outside taxonomic range by ±12 cm

Interpretive Placement

Soil taxonomic properties are important because they provide a means of connecting our conceptual models of soils with real landscape counterparts. Although these concepts are meaningful to a pedologist, soil survey users desire information on how soils will respond relative to given land uses at alternate levels of use and management. Accordingly, taxonomic inclusions to the reference taxa in this study were evaluated to determine which were most influential to soil behavior for the major land uses in the study area. As a guideline, the soil interpretations section (Section 603) of the *National Soils Handbook* (Soil Survey staff, 1983) was used to make comparisons between each of the major soil taxa that were encountered in the transects.

To arrive at logical interpretive groupings, the concepts of similar and dissimilar soils were used. The *National Soils Handbook* (Soil Survey Staff, 1983) defines similar soils as those soils with properties that are marginally outside the limits of the range of properties of the reference taxon. Dissimilar soils are those soils with properties that are significantly outside the limits of the range of properties of the reference taxon. Admittedly, these definitions of similar and dissimilar were somewhat arbitrary and user-biased. We did not address which dissimilar soils might be limiting with respect to specific land uses. To find interpretive compositional purities, the percentages of the reference taxon and soils similar to the reference taxon were combined. Only the dissimilar soils were considered to be inclusions for interpretive compositional analysis. Applications of these definitions were obviously subjective and dependent on intended land use.

Table 11-2 displays the taxonomic properties that were considered interpretatively dissimilar to the reference taxa for each of the map units. Because all units transected were Paleustalfs, dissimilar inclusions at the great group level were the same for all of the units. Below the great group level, the dissimilar inclusions varied from one map unit to the next. Specific inclusions for each taxonomic level are discussed below.

The only inclusions encountered at the order level were Mollisols, which were, in all cases, regarded as similar to the reference taxon because they differed only in the presence of a mollic epipedon that was borderline in color, thickness, and organic C content. At the suborder level, inclusions of Aqualfs were considered dissimilar because of the number of land uses that would be adversely affected by wetness throughout for significant periods.

At the great group level, the Crockett and Spiller soils were classified as Paleustalfs because of an abrupt textural boundary between the surface (A and/or E) and subsurface Bt horizons. In contrast, the Rader and Robco qualified as Paleustalfs, not because they have abrupt textural boundaries to clayey Bt horizons, but because clay decreases in the argillic horizon were less than 20% (relative basis) from the maximum within a vertical distance of 1.5 m of the surface. An abrupt textural boundary between the surface and subsurface horizons restricts water movement and root penetration; observations of abrupt contacts in the Rader and Robco units were thus considered dissimilar. When an abrupt textural boundary was absent in the Crockett and Spiller units, the inclusions were also considered dissimilar.

At the subgroup level, aquic intergrades were not considered dissimilar because these soils are presumably wet for only short periods. Arenic intergrades with thickness error limits of ±12 cm and borderline fine sandy loam textures were not considered dissimilar because most land uses were not affected by variances of this magnitude according to interpretive guidelines. The presence or absence of vertic properties was important for differentiating similar and dissimilar soils at the subgroup level.

At the family level, Crockett and Spiller soils are classified in the fine particle-size family; while Rader and Robco soils are classified as fine-loamy and loamy, respectively. The clay boundary between these particle-size classes is 35%. To allow for error in our field clay percentage estimations, the clay boundary percentages for stated taxonomic ranges were allowed to range ±2%.

At the series level, only inclusions with significant changes in surface thickness (A and/or E horizons) were considered dissimilar. Series surface-thickness ranges were allowed to vary by ±12 cm, which is the same as that allowed at the subgroup level (rationale presented previously). Also, inclusions involving textures of epipedons that were fine sandy loam vs. loamy fine sand were considered similar.

To summarize, the inclusions considered to be dissimilar to Crockett were those soils with aquic soil moisture regimes, nonvertic properties, and fine-loamy control sections (<33% clay); to Spiller, those soils with surface horizons greater than 60 cm or less than 13 cm thick, and fine-loamy control sections (<33% clay); to Rader, those soils with aquic moisture regimes, fine control sections (>37% clay), vertic properties, and surface horizons greater than 100 cm or less than 35 cm thick; and for Robco, those soils with surface horizons greater than 110 cm and less than 35 cm, and clayey control sections (>37% clay).

RESULTS AND DISCUSSION

Taxonomic Composition

The compositional purities of the four reference taxa for which the transected map units were named are shown in Fig. 11-2. The histograms

QUANTIFYING MAP UNIT COMPOSITION

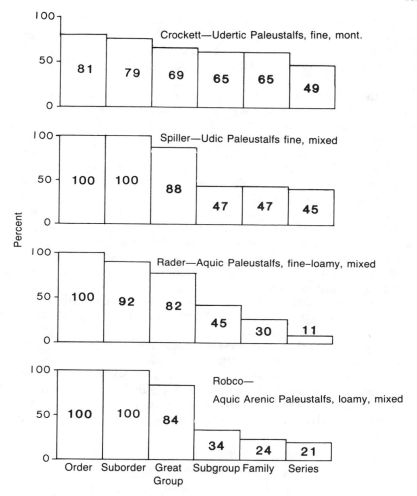

Fig. 11-2. Taxonomic composition of the transected map units (number in each bar represents taxonomic purity).

show how the major taxonomic inclusions occur in relation to different taxonomic levels. The percentage of inclusions increases systematically with categorical specificity as one would expect in a hierarchical classification scheme. More precise statements about soil properties can be made at lower categorical levels. Generally, at the order and suborder levels, taxonomic purity is relatively high (79–100%). Crockett is lowest because of inclusions of soils with mollic epipedons. In addition, a few pedons in the Crockett transects had subsoils dominated by low chroma even though slopes ranged from 3 to 6%.

The first significant decrease in taxonomic purity is at the subgroup level for Spiller, Rader, and Robco map units. At this level, the presence or absence of gray mottles within 75 cm of the surface is used to differentiate udic and aquic intergrades. The presence and location of these mottles varies greatly from pedon to pedon. These mottled soil colors, like those for the Crockett map unit, do not always follow landscape position and may reflect relict gleying conditions.

At the family level, differences in particle-size class account for most of the inclusions in the Rader and Robco mapping units. For example, up to 15% inclusions of soils with fine textures occur in Rader and up to 10% inclusions with loamy textures occur in Robco. The textural variability of these soils seems to lack a predictable pattern and reflects variability in the underlying geologic sediments. Thus, it is difficult to delineate texturally homogeneous units of these soils based solely on soil/landscape models.

The compositional purities are lowest at the series level, although the reference taxon for which each map unit was named was dominant. Purities ranged from a low of 11% for Rader to a high of 49% for Crockett. The major reason for impurity at this categorical level was the relatively narrow range of soil colors permitted by the respective series descriptions. Broadening the color ranges with respect to value and chroma would increase taxonomic purities of these map units.

Figure 11-3 displays the mean taxonomic percentages of the reference taxa at the series level. The mean confidence intervals were ±8% for Spiller and Robco, ±9% for Rader, and ±19% for Crockett, using an 80% probability level. Sufficient observations were obtained for each map unit, except Crockett, to estimate mean values within ±15%. Because of the significant percentage of Mollisol inclusions in some of the Crockett delineations, confidence limits for mean estimates of this soil would be slightly higher than ±15%, with a 80% probability level. Attempts to map these Mollisols separately from the Crockett soils have not been successful because

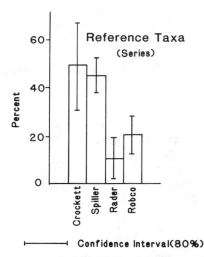

Fig. 11-3. Taxonomic composition and confidence intervals of the transected map units.

of inherent and often short-range soil variability that cannot be delineated at the scale of mapping.

Interpretive Composition

To arrive at interpretive compositional purities, the total similar inclusions were combined with the percentages of their respective taxa (Fig. 11-4). This combination created broader categories that increased interpretive purity considerably. For Crockett and Spiller, the compositional interpretive purity increased to more than 80%. The interpretive composition of the Rader and Robco improved also, but only to 48 and 52%, respectively.

Five transects per each map unit yielded sufficient observations to estimate the interpretive compositional purities within 15% of the mean, at 80% probability level. These data demonstrate that the soil components varied predictably from one delineation to the next within each of the map units selected for this study.

Map-Unit Design

The *National Soils Handbook* (Soil Survey Staff, 1983) recognizes four map unit concepts to accommodate differences in map scale, land use and

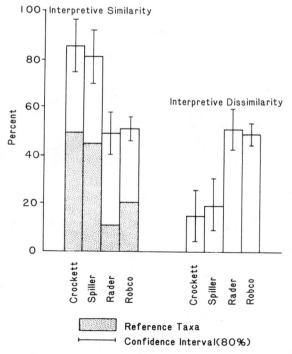

Fig. 11-4. Interpretive composition and confidence intervals of the transected map units.

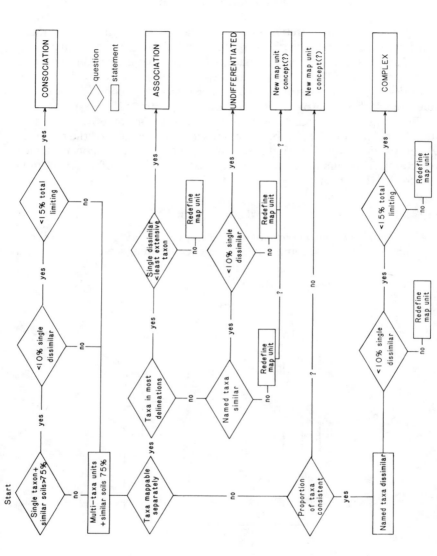

Fig. 11-5. Flow diagram defining the four map unit concepts following the concepts outlined in the *National Soils Handbook* (Soil Survey Staff, 1983).

management, and compositional purities. Map units must account for at least 75% of the observed variability. These map unit design concepts are based on whether a single taxon or multiple taxa are required to account for the variability, and whether the named taxa are similar or dissimilar. The consociation is the only single-taxon unit. The named taxa of the undifferentiated units must be similar; while taxa named in a complex are assumed to be dissimilar. Taxa in an association may be either similar or dissimilar. The named taxa of mapping units may actually constitute much less than 75% of the soil composition as long as these taxa plus similar soils add up to 75%. The aggregate of dissimilar soils is not allowed to exceed 25% in any case. The choice of a mapping unit designation can depend on the definition of similarity. Undifferentiated units, for example, may consist of contrasting soils that are grouped together because a single overriding factor, such as steep slope or extreme stoniness, determines the overall use and management. By definition, such soils are then termed "similar."

We have constructed a flow diagram (Fig. 11-5) that graphically portrays the principal relationships between the four main map unit types. Construction of the flow chart revealed the need for at least one new map unit concept. A counterpart to the complex is needed in cases where the proportion of the named taxa do not occur in a consistent coterminous pattern from delineation to delineation. The interpretive purities of this unit would not change appreciably from the complex.

All four map units in this study were originally designed as single-taxon consociations. The interpretive compositional purities of the Crockett and Spiller units were more than the 75% as required for consociations. In addition, no single dissimilar soil comprised more than 10%. These percentages support the original map unit design.

Rader and Robco both had interpretive purities of about 50%, which is clearly outside the range allowable for consociations. In both of these units, three taxa were needed to attain the 75% level of purity needed for a multitaxa unit. Even though the dominant reference taxon for each of these units occurred in most delineations in consistent proportions, the remaining named components did not. Hence, by definition, these units will not meet the requirements for a complex. As a result, a counterpart to the complex must be used because the named components could not be mapped separately and did not occur in a consistent coterminous pattern from delineation to delineation. Alternatively, the definition of a complex map unit could be redefined to accommodate this conditionality.

Future Trends and Implications for Soil Survey

The results of this study and our field experience lead us to believe that many, if not most, map units should be designed as multitaxa units. It is important to convey to the user that map units of a survey area do not all contain the same level of compositional purity. If soil surveys are to gain user confidence and credibility, they must inform the user that soil varia-

Table 11-3. Map unit composition probability table showing means and confidence intervals of reference taxon and soils similar to the reference taxon.

Map unit	Confidence interval†		
	Low	Mean	High
Crockett	75	86	97
Spiller	70	81	92
Robco	47	52	57
Rader	39	48	57

† There is an 80% probability that the confidence interval presented contains the true population mean for the referenced taxon of each of the transected map units.

Table 11-4. Map unit composition table categorized into high, medium, and low probability classes.

High	Medium	Low
Crockett†	Spiller	Rader
--	--	Robco
--	--	--

† There is an 80% probability that the interpretative composition is >75% for Crockett, 50 to 75% for Spiller, and <50% for Rader and Robco map units.

Table 11-5. Percentage occurrence of soil property compositional data for drainage and surface texture.

Map unit	Drainage (% observed in given classes)				
	Very poorly	Poorly	Somewhat poorly	Moderately well	Well
Crockett	--	--	32	36	21
Spiller	--	--	8	18	74
Rader	--	11	62	22	5
Robco	--	--	7	29	64

Map unit	Surface texture (% observed in given classes)				
	Loamy-sand	Sandy-loam	Loam	Sandy-clay-loam	Clay-loam
Crockett	--	35	58	2	5
Spiller	74	26	--	--	--
Rader	30	65	5	--	--
Robco	93	7	--	--	--

Table 11-6. Central tendency statistics of selected compositional properties.

Map unit	Surface thickness (cm)		Slope (%)	
	Mean	SD†	Mean	SD
Crockett	20	4	1.7	1.0
Spiller	42	15	1.8	0.7
Rader complex	61	25	<1.0	--
Robco complex	68	29	2.1	0.8

† Standard deviation.

bility is inherent, sometimes even in close-range spatial intervals, precluding delineation of certain soil bodies on both large- and small-scale maps.

To be more precise and to further enhance predictability of map unit variability, we can rate the individual map units of a survey area by making probability statements about the means and confidence intervals of the dominant taxon (taxa) and soils similar to the dominant taxon (taxa). A table could be inserted into soil surveys that would either rate the soils in order of absolute variabilities (Table 11-3) or group them into interpretive categories of high, medium and low (Table 11-4). As an example of the former, this rating would allow us to state with 80% assurance that the dominant taxon for a particular map unit makes up 75 to 97% of the area. In the case of interpretive groupings, we could say that the dominant taxon of a particular map unit has a mean occurrence of greater than 75% for the high category, 50 to 75% for the medium category and less than 50% for the low category, again with 80% assurance.

To circumvent problems associated with the concept of similar and dissimilar soils, physical, chemical and landscape compositional soil properties could be summarized and presented in soil surveys without interpretive bias. This approach is shown in Table 11-5. In the Crockett unit, for example, 32% of the transect observations were considered somewhat poorly drained, 36% moderately well-drained and 32% well-drained. Other soil properties that could be analyzed in this fashion include permeability, shrink/swell, and infiltration. Additional properties such as horizon thickness and slope percentage could be quantified by central tendency statistics (Table 11-6). The properties shown in these tables are relevant to no particular land use; the user could scan the table and determine which properties are important for any land use desired.

The current method in soil surveys is to provide soil property data for a typifying pedon of each soil series used in a soil survey area. This method, however, does not provide information about ranges in soil properties or conditions within the delineations of the map unit. Ultimately, the name we place on a soil map unit may be somewhat irrelevant. The important question is how we quantify and portray the composition and variability of the units we map.

SUMMARY AND CONCLUSIONS

The results of this study demonstrate that both taxonomic and interpretive purities of map units in the Tertiary Gulf Coastal Plain of Texas are much less than indicated by soil survey reports. As many as 50% of all map units in soil surveys in this region may need to be designed as multitaxa units. A critical examination of the mapping unit concepts listed in the *National Soils Handbook* (Soil Survey Staff, 1983) reveal that at least one new definition is needed—a complex unit where the named components do not all occur in a consistent coterminous pattern. Alternatively, the definition of a complex could be redefined to waive the requirement that dominant components occur in a "consistent coterminous pattern."

To increase user confidence and soil survey credibility, map unit composition probability tables and soil property compositional tables should be inserted into soil surveys. These tables would permit the user to make land-use decisions based on expectations of finding the named components of a map unit independent of predetermined interpretive biases. The concepts of similar and dissimilar soils are dependent on land use and too often involve subjective (i.e., nonstandardized) judgments by soil survey personnel. Likewise, definitions requiring determination of "limiting" vs. "nonlimiting" dissimilar soils in map units is use-dependent and thus cannot be generalized for a map unit name where multiuse interpretations are made.

Considering the magnitude of map unit variability commonly found in many soil survey areas and the increasingly technical interpretations being made of soil properties, the future trend in soil survey must be one that encompasses more quantification of map unit composition.

REFERENCES

Amos, D.F., and E.P. Whiteside. 1975. Mapping accuracy of a contemporary soil survey in an urbanizing area. Soil Sci. Soc. Am. Proc. 39:937–942.

Arnold, R.W. 1977. CBA-ABC-Clean brush approach achieves better concepts. p. 61–92. *In* Soil survey quality. Proc. New York Conf. Soil Mapping Quality. Bergmanin, NY. 5-7 Dec. 1977. Cornell Univ. Agric. Exp. Stn. in cooperation with the USDA-SCS.

Arnold, R.W. 1981. Binomial confidence limits as estimators of classification accuracy. Agron. Mimeo. 81-7. Cornell Univ., Ithaca, NY.

Bascomb, C.L., and M.G. Jarvis. 1976. Variability in three areas of the Denchworth soil map unit. I. Purity of the map unit and property variability within it. J. Soil Sci. 27:420–437.

Brubaker, S.C. 1989. Evaluating soil variability as related to landscape position using different statistical methods. Ph.D. diss. Texas A&M Univ., College Station (Diss. Abstr. 89-216A6).

Bureau of Economic Geology. 1981. Austin geologic atlas sheet. Univ. Texas, Austin, TX.

Crosson, C.L., and R. Protz. 1974. Quantitative comparison of two closely related soil mapping units. Can. J. Soil Sci. 54:7–14.

Day, P.R. 1965. Particle fractionation and particle-size analysis. p. 545–566. *In* C.A. Black et al. (ed.) Methods of soil analysis. Part 1. Agronomy Monogr. 9. ASA, Madison, WI.

Edmonds, W.J., and M. Lentner. 1986. Statistical evaluation of the taxonomic composition of three soil map units in Virginia. Soil Sci. Soc. Am. J. 50:997–1001.

Klich, I., L.P. Wilding, and A.A. Pfordresher. 1990. Close-interval spatial variability of Udertic Paleustalfs in East-Central Texas. Soil Sci. Soc. Am. J. 54:489–494.

McCormack, D.E., and L.P. Wilding. 1969. Variation of soil properties within mapping units of soils with contrasting substrata in northwestern Ohio. Soil Sci. Soc. Am. Proc. 33:587–593.

Miller, F.P., D.E. McCormack, and J.R. Talbot. 1979. Soil surveys: Review of data collection methodologies, confidence limits and uses. p. 57–65. *In* The mechanics of track support piles and geotechnical data. Transportation Res. Rec. 733. Trans. Res. Board, NAS, Washington, DC.

Mokma, D.L. 1987. Soil variability of five landforms in Michigan. Soil Surv. Land Eval. 7:25–31.

Powell, J.C., and M.E. Springer. 1965. Composition and precision of classification of several mapping units of the Appling, Cecil, and Lloyd series in Walton County, Georgia. Soil Sci. Soc. Am. Proc. 29:454–458.

Ransom, M.D., W.W. Phillips, and E.M. Rutledge. 1981. Suitability for septic tank filter fields and taxonomic composition of three soil mapping units in Arkansas. Soil Sci. Soc. Am. Proc. 45:357–361.

Sellards, E.H., W.S. Adkins, and F.B. Plummer. 1932. The geology of Texas. Univ. Texas Bull. no. 3232.

Soil Survey Staff. 1951. Soil survey manual. USDA Handbk. 18. U.S. Gov. Print. Office, Washington, DC.

Soil Survey Staff. 1967. Soils memorandum 66: Application of the soil classification system in developing or revising series concepts and in naming map units. USDA-SCS.

Soil Survey Staff. 1981. Examination and description of soils in the field. p. 101–105. *In* USDA-SCS soil survey manual. Issue 1 (rev.). U.S. Gov. Print. Office, Washington, DC.

Soil Survey Staff. 1983. National soils handbook. U.S. Gov. Print. Office, Washington, DC.

Soil Survey Staff. 1987. Keys to soil taxonomy. Technical monograph 6, third printing. Dep. of Agronomy, Cornell Univ. Ithaca, NY.

Wilding, L.P. 1988. Improving our understanding of the composition of the soil-landscape. p. 13–35. *In* H.R. Finney (ed.) Proc. Int. Interactive Workshop Soil Resources: Their inventory, analysis and interpretation for use in the 1990's. 22–24 Mar. 1988. Educ. Dev. System, Minnesota Ext. Serv., Univ. Minnesota, St. Paul.

Wilding, L.P., and L.R. Drees. 1983. Spatial variability and pedology. p. 83–113. *In* L.P. Wilding et al. (ed.) Pedogenesis and soil taxonomy. I. Concepts and interactions. Elsevier Sci. Publ. Amsterdam.

Wilding, L.P., R.B. Jones, and G.M. Schafer. 1965. Variation of soil morphological properties within Miami, Celina, and Crosby mapping units in west-central Ohio. Soil Sci. Soc. Am. Proc. 29:711–717.

12 Using Systematic Sampling to Study Regional Variation of a Soil Map Unit

G. W. Schellentrager
USDA-SCS
Des Moines, Iowa

J. A. Doolittle
USDA-SCS
Chester, Pennsylvania

ABSTRACT

Studies were conducted in three New England states to determine the taxonomic homogeneity of a map unit. In New England, extensive areas of Lyman (loamy, mixed, frigid Lithic Haplorthods) and Tunbridge (coarse-loamy, mixed, frigid Typic Haplorthods) soils occur in an intricate pattern on over 300 000 ha of upland area. Commonly these soils are mapped as multitaxonomic map units. However, in areas of Lyman and Tunbridge soils, determination by depth to bedrock is complicated due to coarse fragments and the irregular bedrock surface. Four study sites were selected in delineations thought to represent the central concept of the Tunbridge-Lyman map unit. Systematic sampling using ground-penetrating radar (GPR) identified significant portions of the map unit to have characteristics outside the range of identified taxa. Results revealed that soil scientists are consistently underestimating the depth to bedrock. Soil-bedrock models have been reconstructed in order to improve soil map unit design, descriptions, and interpretations.

Considering the extent of soils, sampling intensities used to support soil survey interpretations are exceedingly small. As they traverse the landscape, soil scientists make a limited number of borings and predict the occurrence of soils across large areas from landforms, vegetative patterns and their associated signatures on aerial photographs. These predictions are based on conceptual models that soil scientists have been taught or have developed

---

Copyright © 1991 Soil Science Society of America, 677 S. Segoe Rd., Madison, WI 53711, USA. *Spatial Variabilities of Soils and Landforms.* SSSA Special Publication no. 28.

after extensive observations of soils as they occur on the landforms. If the working model of the soil scientist is reliable, these predictions are reasonably accurate.

With the constraints imposed by our current technology, costs, and the inherent complexities of some soil patterns, errors in estimating the taxonomic composition of a map unit are unavoidable. Miller et al. (1979) noted three factors contributing to erroneous estimates of map unit composition. First, the predictive value of landscapes is not perfect. Many of the surface features that are used to separate soils in the field are so subtle that even the most skilled soil scientists cannot map them precisely. Some soil boundaries are not expressed by surface features.

Second, traditional sampling intensities for verifying map unit composition are statistically inadequate. They allow reasonable accuracy at a realistic cost only because of the reliability in predictive value of the landforms. This assumes that soil scientists are adequately trained and adept at landform and photograph interpretation.

Third, traditional sampling methods are commonly biased. Soil scientists do not choose sites for the verification of their landform models at random. Soil scientists are aware that less characteristic soils and landforms are present, and may examine some of these areas to get an idea of the variability in the map unit. However, sampling tends to be biased toward the most prominent soils and landforms (Miller et al., 1979). The soil cannot be examined at enough places to insure present sampling methods and intensities are not misleading in some areas.

This study evaluated depths to the soil–bedrock interface, as identified by field soil scientists and characterized in map units of progressive soil surveys. Knowledge of this one diagnostic soil characteristic is critical to soil interpretations and map unit design. Where bedrock occurs within 1.5 m of the soil surface; limitations to rooting depth of trees, percolation of water, and absorption capacity for effluents occur. Additionally, costs of excavation increase dramatically.

In many areas, it is difficult to examine soil profiles and determine the depth of bedrock using tiling shovels, hand augers, and probes because rock fragments or dense soil layers limit penetration. Because of the difficulties experienced in observing the soil–bedrock interface, interpretations concerning soil depth are often based on soil–bedrock–landform relationships and on anticipated rather than confirmed depths to bedrock. In areas of shallow or less variable depths to bedrock, these interpretations are reasonably accurate. Where depth to bedrock is highly variable, interpretations are less accurate and depths are likely to be under- or overestimated. Unfortunately, at the scale at which most detailed soil surveys are conducted, the depth to bedrock often cannot be predicted from landform position alone. Improved soil–landform information is needed in order to provide reliable and accurate soil interpretations.

In glaciated upland areas of the New England province (Thornbury, 1965), major relief features are manifestations of underlying rock formations. In these areas, the till mantle is a thin veneer and soil depth classes are

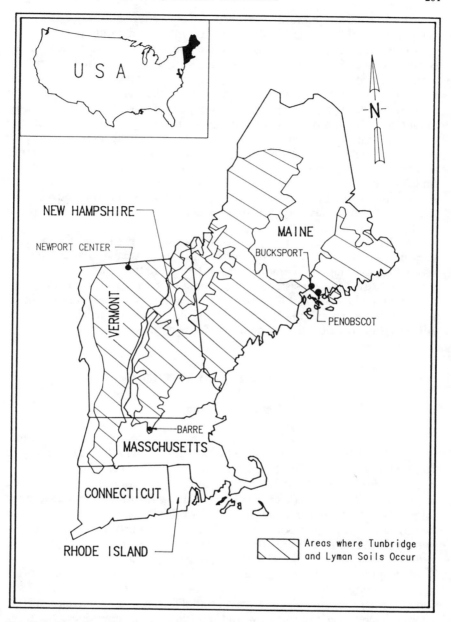

Fig. 12-1. Location of study sites within the New England province.

commonly used as criteria for differentiating soil series. Often, very shallow (<25 cm) soils are of minor extent and are rarely recognized in soil map unit names. Shallow (25–50 cm), moderately deep (50–100 cm), and very deep

(>150 cm) soils are commonly mapped. No soil series have been established for the deep (100-150 cm) soil depth class in these upland areas.

This chapter summarizes and expands on the results of earlier studies (Collins et al., 1989a; Doolittle et al., 1988) that used systematic sampling with GPR in delineations of Tunbridge-Lyman complex on 3 to 8% slopes. Extensive areas of the moderately deep to bedrock soils, Tunbridge, and the shallow to bedrock soils, Lyman, are mapped on more than 300 000 ha of upland areas within the New England province (Fig. 12-1). These soils are commonly mapped as multitaxonomic map units. Map unit descriptions locate the Lyman soils on slightly higher ridgetops and on steep backslopes where bedrock is presumed to be shallow to the surface. Tunbridge soils occur on lower, upper backslopes and on gently sloping areas where the bedrock is presumed to be deeper. Rock fragments in the soil profile and variable depths to bedrock have complicated the accurate mapping and description of these soils.

Objectives of this study were to: (i) assess the predictive accuracy of present soil-bedrock-landform models, (ii) use systematic sampling for estimating soil map unit composition, (iii) estimate the taxonomic composition, and assess the consistency of mapping of areas of Tunbridge-Lyman complex on 3 to 8% slopes in three widely separated regions of New England, and (iv) improve map unit design, descriptions, and interpretations through construction of three-dimensional landform models.

MATERIALS AND METHODS

Systematic Sampling

In a systematic sampling program, the first sampling point is randomly selected; all other points are located at a regular interval, usually in a grid format (Berry, 1962). Compared with simple random sampling, systematic sampling ensures greater areal coverage and information (King, 1969), and is generally more precise (Webster, 1977). However, with systematic sampling all points do not have an equal probability of being selected and errors can be introduced if the periodicity of variations in soil properties corresponds with the interval of the grid (Berry, 1962).

Although suited to soil-landform studies, systematic sampling has had limited application in soil survey operations. Systematic sampling often requires greater expenditure of time, money, and effort per unit of soil sampled than random point or transect sampling methods. Generally, surface probing techniques are too slow and laborious to accommodate the more intensive sampling effort required with systematic sampling. As a result of these limitations, systematic sampling has been restricted principally to research rather than soil survey efforts (Wilding, 1985).

With the advent of GPR and its application to soil-landscape investigations, many of the time and cost constraints imposed by systematic sampling have been overcome. Systematic sampling with GPR has helped to charac-

Table 12-1. Landform parameters at study sites.

Site	Relief	Slope		
		Average	Minimum	Maximum
	m	%		
Bucksport	2.4	4.2	0.5	11.0
Penobscot	2.1	2.9	0.5	6.8
Barre	3.8	5.1	0.5	14.5
Newport Center	1.7	4.4	1.1	9.0

terize soil–landform relationships on glaciated uplands in Maine (Collins et al., 1989a), loess-covered coastal plain sediments in Delaware (Rebertus et al., 1989), and areas of karst in Florida (Collins et al., 1989b). In these studies, an aligned systematic sampling or "checkerboard" pattern (Berry, 1962) was used. This method economizes time by allowing straight, parallel traverses along defined paths. The GPR provides a continuous profile of subsurface features along a traverse, thereby minimizing errors resulting from any periodicity of soil or rock properties. Radar profiles were examined to insure that observation points do not coincide with a recurring soil depth interval.

Sampling grids used to characterize soil–landform relationships using GPR have varied in size from 9 to 10 m (Collins & Doolittle, 1987), with observations made at 1.0-m intervals (110 observation points); to 152 by 152 m (Rebertus et al., 1989), with observations at 3.05-m intervals (1326 observation points).

Field time is related to grid size and interval. Traverses are conducted at speeds of 3 to 8 km h^{-1}.

Study Area

Four study sites were selected in three New England states. The study sites were located near, and named after, the towns of Bucksport and Penobscot, ME; Barre, MA; and Newport Center, VT (Fig. 12-1). Sites occurred on similar landscapes and were dominated by similar appearing landforms. Each site was located on an upper backslope of a broad low hill. Though similar in appearance, all the sites were found to be significantly different from one another ($P = 0.001$ level) in terms of average slope gradient and relief (Table 12-1).

Radar System

The GPR system used was the Subsurface Interface Radar (SIR) System-8.[1] Components used in this study included the Model 4800 control unit, power distribution unit, ADTEK SR 8004H graphic recorder, ADTEK DT 6000 tape recorder, and 120-MHz antenna. Principles of operation have been described by Olson and Doolittle (1985).

---

[1] Manufactured by Geophysical Survey Systems, Inc., 13 Klein Dr., North Salem, NH 03073-0097. Trade names have been used to provide specific information and do not constitute endorsement by the authors.

The accuracy of GPR in determining the soil-bedrock interface has been documented by Collins et al. (1989a) in a related study. In that study a 90-m trench was excavated to bedrock at the Penobscot site to determine the correlation between scaled radar estimates and screw auger measurements of the depths to bedrock, with the actual observed bedrock depths. The average depth to bedrock within the trench was 147.1 cm with a range of 83.8 to 238.8 cm. The average difference in depth to bedrock between the scaled radar imagery and the ground-truth measurements (trench) was 6 cm. However, the average difference in the depth to bedrock between the auger and the ground-truth measurements was 76.5 cm. In addition, 64% of the scaled radar imagery was within 0 to 5 cm of the actual depth to bedrock and 87% was within 0 to 10 cm. This close correlation between scaled radar depths and actual depths are in agreement with those reported by Shih and Doolittle (1984).

Field Procedures

Each study site was selected by soil survey project leaders and was representative of the central concept of the Tunbridge-Lyman complex on 3 to 8% slope map units as it occurred in the respective survey areas. Grids were established on each site and ranged from 111 by 30 m (418 points) to 42 by 21 m (120 points). The size of the grid varied with the size of the soil delineation and the area accessible to the GPR. Observation intervals were 3.05 m. A transit was used to establish grid corners and to determine surface elevations at each grid intersection. Elevations in each plot were not tied to an elevation benchmark; therefore the lowest recorded surface point was taken as the 0.0-m datum.

RESULTS AND DISCUSSION

Computer generated three-dimensional surface net diagrams of the surface (A), thickness of till (B), and bedrock surface (C) in Fig. 12-2 through 12-5 represent the Bucksport, Penobscot, Barre, and Newport Center study sites. These diagrams show that, though the underlying bedrock provides the structural framework for the landforms, the topography of the bedrock surface is more variable than the ground surface. As in other areas of New England, variations in rock type or strength, and glacial quarrying and abrasion have produced a highly irregular bedrock microtopography (Thornbury, 1965).

The till mantle is thin, and with the exception of the Bucksport site (Fig. 12-2B), is not of uniform thickness. Generally, the till mantle thins over bedrock highs and thickens over depressions. The result is a more regular ground surface than bedrock surface. Assuming a southern direction of glacial advance, thickness of till appears to be greater on lee than on stoss slopes of bedrock highs. The depth to bedrock and the thickness of till are largely controlled by the topography of the underlying bedrock surface rather than landform position.

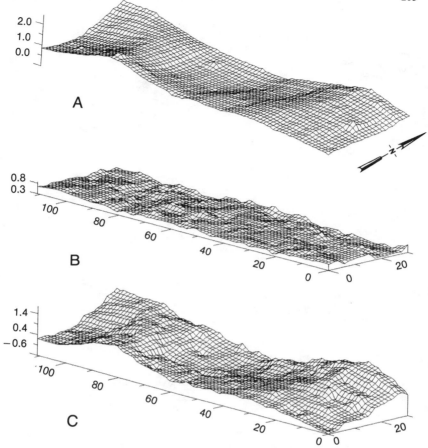

Fig. 12-2. Surface net diagrams of Bucksport, ME study site showing relative soil surface (A), thickness of till (B), and bedrock surface (C). All measurements are in meters. Vertical exaggeration is six times.

Within the study sites, the predictive value of soil-bedrock-slope relationships was weak. Generally, on glaciated upland areas, the bedrock is assumed to be shallow on steep, higher-lying, convex surfaces and to be buried beneath thicker till deposits on lower-lying, concave surfaces. However, little or no correlation existed between depth to bedrock and slope gradient ($r = -0.166$) or depth to bedrock and elevation ($r = 0.033$). In addition, these slope and elevation variables explained very little of the variation ($r^2 = 0.043$) observed in the depth to bedrock within the study areas.

With the exception of the Bucksport site, areas of moderately deep (Tunbridge) and shallow (Lyman) soils were irregular in shape and were more common, but not always dominant, on convex surfaces. These soils were surrounded by larger, and in some places, higher lying areas of deep and very deep soils. While the bedrock forms the framework for the landforms

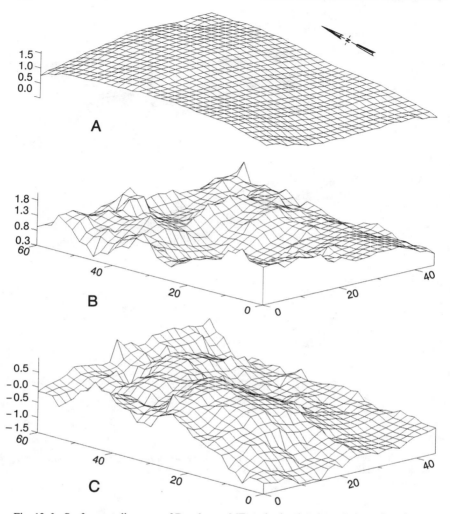

Fig. 12-3. Surface net diagrams of Penobscot, ME study site showing relative soil surface (A), thickness of till (B), and bedrock surface (C). All measurements are in meters. Vertical exaggeration is six times.

and general trends can be inferred, the landform does not have good predictive value for the occurrence of soil depth classes used in soil classification.

The average depths to bedrock were 58, 108, 117, and 96 cm at the Bucksport, Penobscot, Barre, and Newport Center study sites, respectively. With the exception of the Bucksport study site, there was no significant difference in the average depth to bedrock or thickness of till among the study sites. The average soil depth and the taxonomic composition of the Bucksport study site was more representative of the Tunbridge-Lyman complex map unit descriptions in the soil survey reports (Table 12-2A).

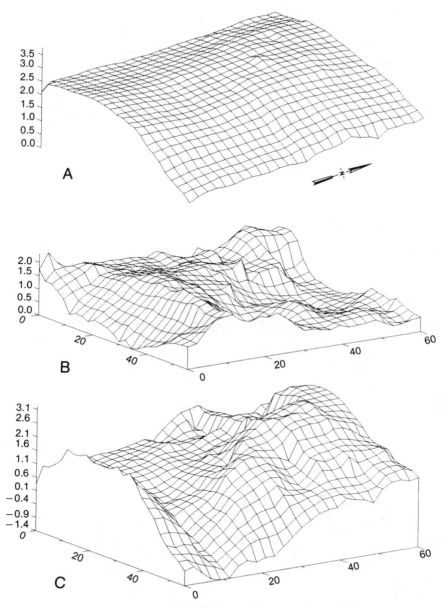

Fig. 12-4. Surface net diagrams of Barre, MA study site showing relative soil surface (A), thickness of till (B), and bedrock surface (C). All measurements are in meters. Vertical exaggeration is six times.

All areas were similarly mapped. Taxonomic compositions determined by traditional transecting methods and described in soil survey reports ranged from 40 to 50% moderately deep (Tunbridge) soils to 30 to 40% shallow

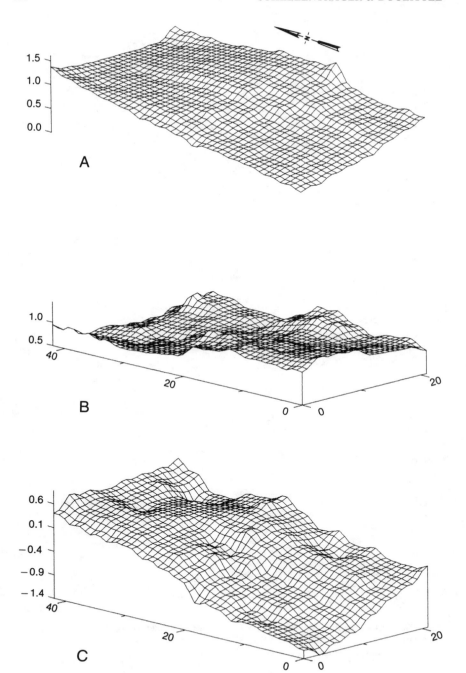

Fig. 12–5. Surface net diagrams of Newport Center, VT study site showing relative soil surface (A), thickness of till (B), and bedrock surface (C). All measurements are in meters. Vertical exaggeration is six times.

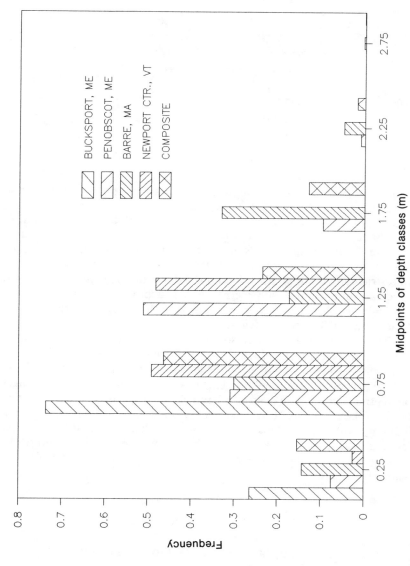

Fig. 12-6. Distribution of depths to bedrock.

Table 12-2. Comparison of depth to bedrock of Tunbridge-Lyman complex, 3 to 8% slopes.

A. Soil survey reports

Study site	Shallow	Moderately deep	Inclusions
		%	
Bucksport	30	50	20
Penobscot	30	50	20
Barre	30	40	30
Newport Center	40	45	15

B. Ground-penetrating radar survey

Study site	Shallow	Moderately deep	Deep	Very deep
		%		
Bucksport	28	72	--	--
Penobscot	9	31	50	10
Barre	14	30	18	38
Newport Center	2	46	52	--
Avg.	13	45	30	12

(Lyman) soils (Table 12-2A). Deep soils, soils varying in drainage class and thickness of the subsoil are commonly described as inclusions. Systematic sampling with the GPR revealed an average composition of 45% moderately deep, 30% deep, 13% shallow, and 12% very deep (Marlow—coarse-loamy, mixed, frigid Typic Haplorthods) soils (Table 12-2B and Fig. 12-6). Based on this limited sample, it appears that traditional surveying procedures are depth restricted (Collins et al., 1989a), biased toward areas of shallow soils, and decrease in accuracy with increasing soil depth (Miller, 1978).

Though very limited in extent compared to the entire occurrence of the map unit, the GPR surveys suggest a need to revise the name of the map unit to Tunbridge–unnamed deep soil–Lyman complex, on 3 to 8% slopes. The unnamed, deep soil occupies 30% of the areas surveyed with the GPR. No soil series has been established for deep soil depth classes in areas of frigid soil temperatures in New England. This is a consequence of recent changes in soil depth concepts. The deep soil depth class, formerly described as being >100 cm to bedrock has been redefined as being 100 to 150 cm to bedrock (Soil Survey Staff, 1981). This change in soil depth criteria and the inability of conventional tools to adequately probe beyond depths of 1.0 m in tills containing large amounts of coarse fragments or dense layers have led to omission of deep soils.

CONCLUSION

Results from this study have revealed that soil scientists underestimated the depth to bedrock in the delineations that they selected as central to the concept of this map unit. A significant proportion of these map unit delineations had characteristics outside the range of the identified taxa, and depending on the land use, contrast considerably in interpretations. This study indicates a need to re-examine soil–bedrock–landform models and to recognize a deep soil series on glaciated upland areas in New England.

The accuracy of soil surveys includes not only the accuracy of the soil maps, but also the accuracy of the map unit descriptions and the validity of their names measured against the standards established for nomenclature. Conventions for naming map units need to fit the natural landforms. One of the easiest ways to improve the quality and accuracy of soil survey information is to use a soil descriptive legend that accurately reflects the natural variability of the soil landform. This implies that we know what the mappable landforms contain. In most cases, however, we do not know this in quantitative terms. If mappable landforms are mixtures of soil taxa, we must say so by designing map units that accurately reflect the variability observed. When studies are undertaken to quantify map unit variability the relative proportions of multitaxonomic map units on identification legends increase at the expense of consociations (Schellentrager et al., 1988).

The quality of a soil survey is best expressed as the degree to which the user gains an understanding of the three-dimensional landscape (Brown, 1985). Systematic sampling using GPR can enhance, document, and promote our understanding of soils and the landscapes on which they occur.

ACKNOWLEDGMENTS

The authors wish to express their appreciation to Donald Clark, Robert Joslin (retired), David Schmidt, Eric Swenson, Bill Taylor, Robert Long, Patrick Berry, and Martha Haynes of the SCS for assisting in the field work for this study. The authors also wish to express their appreciation to Douglas Barnes, SCS, for assisting in the cartographic preparation.

REFERENCES

Berry, B.J.L. 1962. Sampling, coding and storing flood plain data. USDA Agric. Handb. 237, U.S. Gov. Print. Office, Washington, DC.

Brown, R.B. 1985. The need for continuing update of soil surveys. Proc. Soil Crop Sci. Soc. Fla. 44:90–93.

Collins, M.E., and J.A. Doolittle. 1987. Using ground-penetrating radar to study soil microvariability. Soil Sci. Soc. Am. J. 51:491–493.

Collins, M.E., J.A. Doolittle, and R.V. Rourke. 1989a. Mapping depth to bedrock on a glaciated landscape with ground-penetrating radar. Soil Sci. Soc. Am. J. 53:1086–1812.

Collins, M.E., W.E. Puckett, G.W. Schellentrager, and N.A. Yust. 1990. Using GPR for micro-analysis of soils and karst features on the Chiefland Limestone Plain in Florida. Geoderma 47:159–170.

Doolittle, J.A., R.A. Rebertus, G.B. Jordan, E.I. Swenson, and W.H. Taylor. 1988. Improving soil-landscape models by systematic sampling with ground-penetrating radar. Soil Surv. Horiz. 29: 46–54.

King, L.J. 1969. Statistical analysis of geography. Prentice-Hall, Englewood Cliffs, NJ.

Miller, F.P. 1978. Soil survey under pressure: The Maryland experience. J. Soil Water Conserv. 33:104–111.

Miller, F.P., D.E. McCormack, and J.R. Talbot. 1979. Soil surveys: Review of data collection methodologies, confidence limts, and uses. Transp. Res. Rec. 733. NAS, Transp. Res. Rec. Board, Washington, DC. p. 57–65.

Olson, C.G., and J.A. Doolittle. 1985. Geophysical techniques for reconnaissance investigations of soils and surficial deposits in mountainous terrain. Soil Sci. Soc. Am. J. 49:1490–1498.

Rebertus, R.A., J.A. Doolittle, and R.L. Hall. 1989. Landform and stratigraphic influences on variability of loess thickness in northern Delaware. Soil Sci. Soc. Am. J. 53:843–847.

Schellentrager, G.W., J.A. Doolittle, T.E. Calhoun, and C.A. Wettstein. 1988. Using ground-penetrating radar to update soil survey information. Soil Sci. Soc. Am. J. 52:746–752.

Shih, S.F., and J.A. Doolittle. 1984. Using radar to investigate organic soil thickness in the Florida Everglades. Soil Sci. Soc. Am. J. 48:651–656.

Soil Survey Staff. 1981. Soil survey manual. USDA-SCS, U.S. Gov. Print. Office, Washington, DC. (In press.)

Thornbury, W.D. 1965. Regional geomorphology of the United States. John Wiley & Sons, New York.

Webster, L.P. 1977. Quantitative and numerical methods in soil classification and survey. Clarendon Press, Oxford, United Kingdom.

Wilding, L.P. 1985. Spatial variability: Its documentation, accommodation and implications to soil surveys. p. 166–189. In D.R. Nielsen and J. Bouma (ed.) Soil spatial variability. Proc. Workshop ISSS and SSSA, Las Vegas, NV. 30 Nov.–1 Dec. 1984. PUDOC, Wageningen, Netherlands.

13 Confidence Intervals for Soil Properties within Map Units

Fred J. Young

USDA-SCS
Columbia, Missouri

J. M. Maatta and R. David Hammer

University of Missouri
Columbia, Missouri

ABSTRACT

Estimates of the means of various soil properties within map units may be important to users. It seems appropriate to measure and report confidence intervals for the means of selected soil properties within map units. Such properties should be important to the expected use and management of the map unit, or should provide important baseline data for future uses. Data on pertinent soil properties can be collected as part of a systematic method of randomly selected transects, such as is presently used to determine map unit composition in some soil surveys. Standard statistical procedures can then be used to calculate confidence intervals for the means of these properties. The resulting ranges may or may not be wholly within the bounds of the conceptual series or taxonomic range. The ranges can be included in the soil survey report in a table, as ranges in the map unit descriptions, and as part of the database for a digitized soil survey. Quantified ranges for the means of soil properties within map units will significantly improve the quality and usefulness of National Cooperative Soil Survey (NCSS) soil surveys.

A primary purpose for making soil maps is to delineate meaningful map units from the virtually continuous variability of soils on the landscape. The most meaningful and understandable map units for users of soil surveys are those that minimize internal heterogeneity (Webster & Beckett, 1968; Butler, 1980). A map unit description should convey the central tendencies of the soil properties within that unit, and convey the degree of variability around those central tendencies.

Within the NCSS the central tendencies of map units are represented by one or more classes as defined by *Soil Taxonomy* (Soil Survey Staff, 1975).

Typical pedons are used to represent these classes. Variance beyond class limits is described as inclusions.

It is widely recognized that taxonomic classes do not adequately define soil variability within a map unit. Natural distributions of soil properties do not lend themselves to hierarchical arrangements (Webster, 1968). Hole and Campbell (1985) discussed two different, but not mutually exclusive, approaches to soil geography. One approach is based upon idealized taxonomic or genetic units and the other approach upon soil bodies as areal entities with characteristic size, shape, and internal variability. Knox (1965) distinguished between the taxonomic and landscape concepts of soil individuals and noted the difficulty of deriving a system of classification based upon soil landscape individuals. The *Soil Survey Manual* (Soil Conservation Service, 1980) and the *National Soils Handbook* (Soil Conservation Service, 1983) state that taxonomic classes rarely coincide precisely with mappable landscape units.

Outlying soil properties within the map unit are recognized and described as inclusions. Numerous studies have shown that the taxonomic class of the typical pedon underestimates the variability within the map unit even when inclusions are considered (Powell & Springer, 1965; Wilding et al., 1965; McCormack & Wilding, 1969; Edmonds et al., 1982; Edmonds et al., 1985a,b; Edmonds & Lentner, 1986). Much of the map unit not within taxonomic bounds can be considered as "similar inclusions" (Soil Conservation Service, 1980), which are similar in use and management to the named taxon.

At present, NCSS policy is to maintain map units and taxonomic units as separate entities within the soil survey report. Taxonomic units are represented by profile descriptions. The variability within taxonomic units is described with a range of characteristics. Map units, which are designed for the nontechnical user, include a brief synopsis of the typical pedon. Spatial variability is described primarily in terms of limiting (dissimilar) inclusions. Similar inclusions are mentioned briefly if at all. Soil property ranges are seldom described in the map unit description; by policy, the range of characteristics is expressed only in the taxonomic unit description. The total variability of the map unit is represented by some part of the range in characteristics of the taxonomic unit (not included in the map unit description), by dissimilar inclusions that are named and described, and by similar inclusions that are not named and are, at best, briefly described.

How does the soil survey user determine mean or modal values for soil properties of interest within a map unit? Some may erroneously assume that the typical pedon conveys this information. It seems unlikely that any single pedon can accurately represent central values for all ranges of soil properties (Soil Conservation Service, 1983). So the typical pedon is meant only to represent the dominant taxonomic class of the mapping unit. Mean values for soil properties within map units are not explicitly stated in NCSS soil surveys. Ranges for some soil properties, bound by taxonomic limits, can be found in various tables and in taxonomic unit descriptions. But there is no indication that midpoints of these ranges estimate means for specific map units.

How do soil survey users perceive and interpret variability within map units? Jansen and Arnold (1976) noted that soil survey users treat soil series (taxonomic) ranges as cartographic (map unit) ranges. Problems accompany this approximation. The taxonomic range of characteristics is not specific relative to the map unit and for some properties may be wider than the actual range within the map unit (Soil Conservation Service, 1980), although Hudson (1980) disputes this. For other properties, the taxonomic range truncates the natural range within a map unit. Distinguishing these portions of the map unit as "similar inclusions" seems contrived and confusing, particularly to nontechnical soil survey users. The range of characteristics, as part of the taxonomic unit, is a concept and is not designed to approximate a specific map unit.

Many workers have recognized and addressed this problem. Knox (1987) suggested the concept of a "cartographic series" that would be centered on the taxonomic series but would be allowed to encompass the actual range of soil properties within the cartographic unit. Edmonds et al. (1985a) suggested that map unit descriptions be based upon the measured map unit composition rather than on ranges in soil characteristics defined by *Soil Taxonomy* and stated that "attempts to make ranges in soil properties in map units correspond to ranges defined for official soil series are pointless . . ." (Edmonds et al., 1985b). Edmonds and Lentner (1986) believed that soil surveyors should be allowed to describe variability of soil properties as ranges in characteristics for map units. Jansen and Arnold (1976) defined ranges in soil properties for conceptual soil landscape units and noted the disparities between measured ranges for specific landscape units and conceptual ranges for taxonomic units.

The objective of this paper is to propose methodology for determining the mean and variability of soil properties within map units, and to show how this information can be clearly conveyed to the soil survey user. In developing this methodology, the following criteria were applied: (i) the methodology must be relatively simple and easy to apply; (ii) the methodology should build upon established, accepted NCSS policies and procedures; (iii) the results should be as objective and unbiased as possible; and (iv) the resulting ranges should be free from the artificial constraints of a hierarchical taxonomic system.

RANGES DEFINED

The true mean of a soil property within a map unit cannot be determined, so it would be misleading to approximate the mean with a single value. Therefore a range of the mean should be defined; that is, a range of values within which one can have a certain degree of confidence that the true mean lies. Ranges of the mean for a large variety of properties could be determined. Soil survey managers should carefully consider which properties are appropriate for data collection. Considerations should include the expected use and management of the map unit, the types of baseline data that may be necessary for future uses, and the time and expense of collecting data for specific properties. Properties appropriate for data collection might include thickness of the A horizon, depth to some restrictive feature such as a lithic contact, an argillic horizon, a petrocalcic horizon, a fragipan, gley colors, or other important characteristics. Ranges in clay, coarse fragments, pH, salts, Al saturation, and other physical and chemical properties may also be appropriate. Any property requiring laboratory analysis will substantially increase the time and expense associated with data collection. Data collection for some important soil properties, such as hydraulic conductivity or available water capacity, may be beyond the scope of a production soil survey. Other soil properties, such as some that are commonly listed in the taxonomic range of characteristics, are of limited value to most soil survey users and need not be rigorously defined for map units. Soil survey managers can determine specific properties of importance for each map unit, perhaps at the soil survey initial review.

Another important consideration is if the ranges should represent dissimilar inclusions as well as areas of the named taxonomic component and similar soils. A soil survey user may be able to identify these areas in the field, and may be able to avoid them or to manage them differently. For such a user, dissimilar inclusions should not be included within the ranges, but should be clearly identified and located in the map unit. On the other hand, if dissimilar soils are excluded from the database, the resulting ranges do not represent the entire cartographic unit. This complication makes the soil survey more difficult to understand for the nontechnical user. Jansen and Arnold (1976) noted the subjectivity involved in distinguishing between similar and dissimilar inclusions. Objective criteria are needed to make this distinction, and these criteria should be determined by soil survey managers for the soils in the survey area. The ranges developed in this paper represent only the major interpretive component(s) of the map unit.

STATISTICAL CONSIDERATIONS

The easiest method of defining ranges of means is by an extension of the subjective approach currently used to identify typical pedons and define series ranges. This approach is not tenable for cartographic mean ranges due to the strong bias imposed by taxonomic categories and by the necessarily undefined boundaries of the cartographic range. In keeping with an overall trend toward more rigorous soil survey techniques (Reybold & Petersen, 1987) and in the belief that intuitive, subjective bias should be minimized in the soil survey process (Young, 1988), statistical techniques should be used to establish ranges. Confidence intervals have been used for this purpose by De Gruijter and Marsman (1985). Steers and Hajek (1979) used confidence intervals within the NCSS to quantify map unit composition. Confidence intervals for components of map units are currently published in NCSS reports in Florida (Schellentrager et al., 1988).

Confidence interval establishment requires an estimate of variance, which in turn requires random, independent, normally distributed observations.

Data have been collected in a number of ways for soil variability studies. Randomly selected points within map unit delineations have been used (Wilding et al., 1965; McCormack & Wilding, 1969; Bascomb & Jarvis, 1976; Ragg & Henderson, 1980; Ransom et al., 1981; Edmonds et al., 1982; Edmonds et al., 1985a,b; Edmonds & Lentner, 1986). This method is advantageous in that variance for individual pedons can be established. However, De Gruijter and Marsman (1985) point out that locating large numbers of randomly selected points is difficult and time consuming. Random sampling of pedons does not seem feasible for production soil survey work.

USE OF TRANSECTS

The Point Transect Method

Transects have been used to assess soil variability in research (Powell & Springer, 1965; Wang, 1982) and soil survey (Steers & Hajek, 1979; Bigler & Liudahl, 1984; De Gruijter & Marsman, 1985). Transecting techniques are widely accepted within the NCSS, the negative comments of White (1966) notwithstanding. Some degree of transecting is strongly encouraged or required in virtually all NCSS soil surveys.

The point transect seems to be the logical method for data collection. Although a variety of information can be gleaned from transect data, attention within the NCSS has been focused almost entirely on statistical analysis for map unit composition (e.g., Steers & Hajek, 1979; Bigler & Liudahl, 1984). Analysis of transect data is NCSS policy in a few states.

The data necessary to develop confidence intervals for the means of important soil properties can be collected as part of the point observation to determine the soil component. If a profile observation is made, most important soil morphological properties are observed, often as part of the process of classifying the soil. For soil surveys transected for statistical analysis, data collection on most soil properties poses little or no additional time or expense. Properties that require more detailed field observation or laboratory analyses will add to the expense of the survey. Office time will be required for tabulation and calculation but can be minimized with well-designed transect forms and computer programs.

Most soil surveys, however, are not presently transected for statistical analysis. In these surveys, transecting will increase time and expense. The ability to build confidence intervals for soil properties as well as map unit composition, and the need to establish consistent standards for soil survey quality, should serve as impetus for a NCSS policy of transecting for statistical analysis.

Transect Selection

To our knowledge, in most soil surveys transected for statistical analysis, transects to be run are not randomly selected (inferred from Bigler & Liudahl,

1984; Florida SCS soils staff, unpublished data). This biased sampling may provide acceptably accurate results. Most soil survey managers are skilled at recognizing "typical" or representative delineations. However, the population sampled is only those delineations of a map unit that are perceived by the worker as typical; atypical delineations have no chance of being represented in the data. This practice results in a tendency for narrower confidence limits than are truly representative of the population. Counteracting this may be a tendency to select delineations with obvious or extreme features that the soil scientist believes should be included in the data. Excessively uniform delineations may deliberately not be selected, which results in artifically wide ranges.

Biased sampling also introduces the soil scientist as a significant variable. Even skilled, experienced soil surveyors may have significantly different perceptions of what is typical and how to select transects to describe their perceptions. Standard deviations and confidence intervals imply certain assumptions and an objective, scientific basis. Application of statistical techniques to samples selected with no elements of randomness is not scientifically acceptable.

Ideally, each delineation of a map unit should have an equal probability of selection for transect analysis, or, as in the technique of De Gruijter and Marsman (1985), probability in proportion to size. This selection method will insure that transects are random and independent and validates the use of statistical techniques. Statistical validity remains somewhat problematic during a progressive soil survey, because not all delineations will have been identified. It is desirable to conduct transect sampling concurrently with mapping to distribute the transect workload throughout the period of the survey. This distribution will allow the information obtained by transecting to be used during the soil survey process.

Stratification of Survey Area

Stratification of the survey area can address the problem of transect location, increase the efficiency of sampling (Webster, 1977), and identify geographic trends in map unit variability. Stratification involves separating the survey area into subdivisions and sampling each subdivision. Stratification can be accomplished with a grid, such as townships or field sheets or, perhaps, by a preliminary general soil map. As mapping is completed in a subdivision, transects can be selected with all delineations in the subdivision having equal probability of selection. The transects can then be conducted, even though other subdivisions of the survey area are not mapped. At the end of the survey, additional transects (that were previously selected) may be conducted to achieve equal probability between subdivisions.

Disadvantages accompany stratified sampling. Stratification increases the complexity of data collection, record keeping, and analysis. Oversampling some map units could result in inefficient use of field time. Undersampling can be avoided by selecting more transects than are implemented; this also provides some potential transects in reserve for subsequent sampling.

The small numbers of transects required within each subdivision will probably result in unequal probability of selection between subdivisions. Also, the data for a map unit within a subdivision may not be robust and could be distorted by a few aberrant observations.

Transect Weight

Another consideration in transecting methodology is the relationship between transects, delineations, and areas. Wang (1982) showed that some soil properties had higher means in large delineations, although variances were equal for all properties studied. If each delineation is represented by a single potential transect, then all transects will not have equal weight because delineations differ in size. A weighting factor is necessary in the calculations. If transects do not all have equal numbers of observations, which seems likely, an additional weighting factor becomes necessary. De Gruijter and Marsman (1985) eliminated the weighting factor by making probability of selection proportional to the size of the delineation. Steers and Hajek (1979) located potential transects by area, with each area being equal to the expected average size of the management plot. For progressive soil surveys, it seems most appropriate for the mapper to locate potential transects that represent roughly equal areas. Larger delineations may contain several potential transects. Small delineations may need to be grouped. For example, three small delineations may be represented by a single transect, which will be the sum of the three subtransects. Field sheet overlays and a database can be used to keep track of the potential transects.

Locating Transects

Transects can be subjectively placed in the delineation to include as much of the natural landscape variability as possible (Arnold, 1977; Steers & Hajek, 1979; Wilding & Drees, 1983). Such transects could be a single line at right angles to the drainage pattern. Two perpendicular lines might be used for all delineations on more complex landscapes (De Gruijter & Marsman, 1985). Subjective placement and orientation of potential transects may introduce an unacceptable level of bias into the process, and may overestimate the degree of variability. An unbiased approach would locate transects randomly. However, soil properties are not distributed randomly. Experienced soil scientists can perceive the factors that control this distribution. It seems most efficient to utilize the skills of the soil survey workers in transect placement.

DATA ANALYSIS

Mean and Variance

Steers and Hajek (1979) used one-way analysis of variance to determine variance from the transect data. The techniques of Cochran (1977) and

Mendenhall et al. (1971), as used by De Gruijter and Marsman (1985), may be more appropriate. Arnold (1977) included a weighting factor that is used to analyze transect data for map unit composition in some NCSS soil surveys (Bigler & Liudahl, 1984). A position paper "Map unit transects and statistics" has been circulated within the SCS (B. Hudson, 1989, personal communication) that suggests a binomial approach to analysis for map unit composition. A binomial approach is not appropriate for the continuous variables represented by soil properties.

We suggest the following variance formulas, in which n = the number of transects

N = the total number of observations in all transects
x_j = an individual observation
$\bar{y}_i$ = the mean of transect i = the sum of all x_j observations within transect i, divided by the number of appropriate observations (excluding dissimilar inclusions), and
$\hat{u}$ = the overall mean (the mean of all N observations), where

$$\hat{u} = \Sigma x_j / N. \qquad [1]$$

The variance of $\hat{u}$ then is

$$\Sigma(\bar{y}_i - \hat{u})^2 / n(n-1), \qquad [2]$$

and the standard error is the square root of the variance.

This variance formula gives equal weight to each transect. Transects will have unequal numbers of observations due to differing mapping unit dimensions and possible elimination of dissimilar inclusions from the transect data. Arnold's (1977) weighting factor is appropriate even though transects are assumed to represent equal areas. The estimate of the mean should improve with increasing sample size and result in increased confidence in the accuracy of the mean. Means based on more observations are weighted more heavily than are means based on fewer observations.

The weight calculations are as used by Arnold (1977) but are adapted to the formulas of Mendenhall et al. (1971) and Cochran (1977) for systematic sampling

w_i = weight of transect i = number of observations in transect i/average number of observations per transect (N/n).

The weighted variance is then

$$\Sigma[(\bar{y}_i - \hat{u})^2 (w_i^2)] / n(n-1). \qquad [3]$$

Once the overall mean ($\hat{u}$) and standard error (se) have been derived, a confidence interval can be derived with the formula

$$\hat{u} \pm t \times se \qquad [4]$$

CONFIDENCE INTERVALS FOR SOIL PROPERTIES

Table 13-1. An example of confidence interval calculations for surface layer thickness within a map unit of Leonard silty clay loam, eroded, 2 to 4% slopes, in the Audrain County, MO, soil survey area.

Transect number	1	2	3	4	5	6	7
Observed thickness in cm	20	20	†	23	20	15	8
	20	20	40	8	23	8	10
	23	20	23	10	15	5	8
	20	28	23	23	18	5	20
	18	28	20	13	15	25	†
	15	20	20	†	†	10	8
	†	18	25	†	23	18	10
	†	†	23	†	20	8	13
	18	20	20	18	15	10	25
	13			13	15	5	10
Transect sums	147	174	194	108	164	109	112
Transect means, $\bar{y}_i$	18.38	21.75	24.25	15.43	18.22	10.9	12.44
Transect weight, w_i^2	0.9	0.9	0.9	0.69	1.14	1.41	1.14

N = total number of observations = 59.
Overall mean = $\Sigma x_i/N$ = 17.08 cm; (147 + 174 + 194 + 108 + 164 + 109 + 112)/59.
Mean number of observations per transect = 8.43 (59/7).
Variance = $\Sigma(\bar{y}_i - \hat{\mu})^2 w_i^2/n(n-1)$ = 3.5505.
Variance = $[(18.39 - 17.08)^2(0.9)] + [(21.75 - 17.08)^2(0.9)]$ + etc./7(7 - 1).
Standard error = $(v)^{1/2}$ = 1.8843.
80% confidence interval = 17.08 ± t × 1.8843 = 14 to 20 cm; where t = 1.440 for 6 df.
90% confidence interval = 17.08 ± t × 1.8843 = 13 to 21 cm; where t = 1.943 for 6 df.

† Denotes observations of dissimilar inclusions

where t = Student's t at the desired level of confidence, with degrees of freedom = $n - 1$.

Table 13-1 is an example of the calculations for Ap horizon thickness in a map unit of Leonard silty clay loam, eroded, 2 to 4% slopes. (The Leonard series is classified as fine, montmorillonitic, mesic, sloping Vertic Ochraqualfs.) Data were taken from the progressive soil survey of Audrain County, MO. Data are presented for illustrative purposes only. Transects were not randomly selected and do not represent an unbiased sample of the population.

Variance is of transect means, not individual observations. Means are used for several reasons. Point observations of a soil property within a delineation are probably not normally distributed. The specific property will vary with the expression of the factors of soil formation that influence it. However, the means of many samples from a nonnormal population will approximate a normal distribution. Sufficient transects should be conducted so that the means of those properties sampled by transects will approximate a normal distribution.

The variance within each transect is not used in the calculations. Determining this variance is problematic because observations within a transect may not be independent. Positive autocorrelation probably exists within a transect, so adjacent observations should be more correlated than distant observations. Many studies have demonstrated these relationships (e.g., Campbell, 1978; Lanyon & Hall, 1981). Point spacing is an important determinant affecting the quality of the mean and the efficiency of the transect for the calculations presented here. If intervals are too close, observations will be duplicated; but if intervals are too distant, important variability may be missed.

If the survey area is stratified as suggested, then the mean and variance are as follows (adapted from Cochran, 1977, as suggested by De Gruijter & Marsman, 1985)

P = number of potential transects of the map unit
p_j = number of potential transects of the map unit in stratum j
$\hat{u}_{st}$ = overall mean, as determined by stratified sampling.

$$\hat{u}_{st} = \Sigma(p_j\hat{u}_j/P) \qquad [5]$$

where $\hat{u}_j$ is the mean for stratum j. The variance by stratified sampling is

$$V_{st} = \Sigma w_j^2 v_j \qquad [6]$$

where

v_j = variance of stratum j, as calculated by Formula [3]

$$w_j = \text{weight of stratum } j = p_j/P. \qquad [7]$$

One could set P = total area of the map unit; and p_j = area of the map unit in stratum j. Potential transects were used as weighting criteria for consistency with the previously discussed method of defining potential transects and determining variance.

The Confidence Interval and the Tolerance Interval

The confidence intervals developed here are ranges for the mean of a soil property, not ranges of individual observations. For example, if many sampling programs are conducted on the soil property shown in Table 13-1, and means are calculated from each sampling program, then 90% of those means will fall between 13 and 21 cm. Users and soil survey managers would undoubtedly like to have a range for individual observations. For example, what is the surface layer thickness range for 90% of the pedons within that mapping unit?

This range is easily defined from the transect data by treating all observations as independent and developing a histogram (Fig. 13-1). Outliers can be subjectively eliminated and a range defined based on the data. These are the "subjective limits" of Jansen and Arnold (1976). Roughly 90% of the observations are between 8 and 25 cm (Fig. 13-1).

If statistical techniques are used, a slightly wider range is defined. The mean for the data in Table 13-1 and Fig. 13-1 is 17.1 cm. Assuming independence among the observations, the standard deviation (SD) is 6.74. Roughly

CONFIDENCE INTERVALS FOR SOIL PROPERTIES

90% of the observations will fall in the interval 17.1 ± 1.67 × 6.74, where 1.67 is the Student's t value for the 0.90 percentile with 60 degrees of freedom. The resulting 90% range is 5.8 to 28.4 cm. This was one approach examined by Protz et al. (1968) in defining modal profile ranges.

However, this method assumes that the sample mean is the same as the true mean. The standard error (SE) of the sample mean (treating each observation as independent) is $SD/(N)^{1/2}$, or $6.74/7.68 = 0.877$. For this example, we can have a confidence level of 90% that the true mean is between 17.1 ± 1.67 × 0.877, or between 15.6 and 18.6 cm. The 90% range for individual observations must account for this uncertainty about the mean and, consequently, will be wider than the sample range. The appropriate statistic is a tolerance interval, as described by Bowker (1947) and applied by Jansen and Arnold (1976) for this purpose. The tolerance interval is in the form

$$\hat{u} \pm K \times (SD) \tag{8}$$

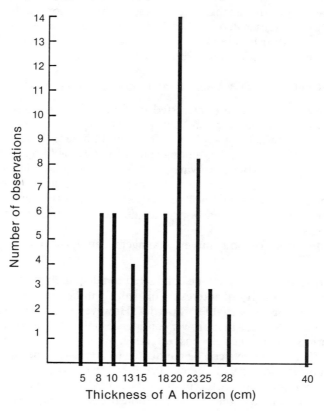

Fig. 13-1. Histogram of observations from transect data in Table 13-1.

where $\hat{u}$ is the sample mean; SD is the sample standard deviation; and the value of K is dependent upon the sample size (n), the confidence level chosen, and the portion of the population to be included.

The K value for the example above is 1.89 for 90% of the population at a 90% confidence level based on a sample size of 59 (Bowker, 1947). The calculated tolerance interval is 4.3 to 29.8 cm. We can have a 90% confidence level that 90% of the pedons within the map unit (excluding dissimilar soils) will fall within that range.

The appropriate use of transect data is to calculate transect means, from which we compute the SE of the mean. A tolerance interval [8] would utilize a K value for $n - 1$ degrees of freedom; where n = number of transects, not the number of individual observations. The SD is that of the transect means, or $[\Sigma(\bar{y}_i - \hat{u})^2/n - 1]^{1/2}$. The resulting wide tolerance interval will tell us about individual transect means, not individual pedons.

Although developing histograms of data can be useful to soil survey managers, the use of statistical techniques to interpret these data is questionable for the following reasons: (i) a tolerance interval can be developed, but its validity is suspect for analyzing transect data because the underlying assumptions of independence have not been fully met; (ii) the individual soil observations, unlike the means, may not approximate a normal distribution (Protz et al., 1968; Jansen & Arnold, 1976). Although a soil property probably exhibits a normal distribution over a broad geographic region, normality may not exist within a specific geomorphic surface.

Coefficient of Variability and Number of Transects

Soil survey managers are interested in the number of transects necessary to adequately define a given map unit. Arnold (1977) discussed techniques to this end, which are applied by Bigler and Liudahl (1984) to a NCSS soil survey. Variance will differ among soil properties (Wilding et al., 1965; Ameyan, 1986). If coefficient of variability (CV) is calculated from transect data on two or more characteristics of a map unit (i.e., map unit composition and one or more important soil properties), it follows that differing numbers of transects will be necessary to define the differing map unit characteristics.

This problem for soil survey managers can have several possible solutions:

1. Sufficient transects can be conducted to adequately define the most variable property of interest. This work will be inefficient in terms of sampling less variable properties. However, the increased accuracy can be reflected in narrower confidence limits. The additional information may not be fully cost effective, but it is certainly not wasted.
2. Sufficient transects can be conducted to adequately define the most important property of interest. Presumably, this property will be map unit composition. Unacceptably wide confidence intervals may result for more variable soil properties. Some properties may be too variable for statistical definition within time and cost constraints of the soil

survey. Subjective estimates for these properties may be appropriate if they are clearly identified.

CARTOGRAPHIC RANGES

Implications of Defining Cartographic Ranges

The ranges of some soil properties within a map unit, as determined by the transecting techniques discussed here, might exceed the limits of the taxonomic unit for which the map unit is named (Jansen & Arnold, 1976). In some cases additional data can be collected to reduce the variance, or confidence limits can be defined for a lower level of probability (e.g., 75 instead of 85%). A multitaxa map unit name may be more appropriate. We suggest that the interests of the user are best served by explicitly stating the actual, calculated range, at a level of accuracy appropriate to the survey area, regardless of taxonomic boundaries. If policy prevents soil survey managers from stating the range, then transecting for statistical significance is of dubious value.

Expressing ranges for properties in multitaxa map units is problematic. Should a range be reported for each component, or should a single range describe all components of the map unit? The proper choice probably depends on the nature of the soil complex and the expected uses of the map unit. A mean value for a complex of two extreme soils may be meaningless; for example, clay content for a complex of coarse-loamy and clayey alluvial soils. On the other hand, it may be difficult and somewhat contrived to report separate ranges for soils that are a continuum, for example, intricately intermixed moderately deep and deep soils. If soil survey users can recognize the components in the field and manage them separately, then separate ranges are probably most useful. If not, a single range may be most appropriate.

Publication of Soil Property Ranges

Means and ranges of soil properties should be published in the soil survey report. Ranges should be expressed in simple, nontechnical terms in the map unit descriptions, and included in more technical terms as a table of ranges of important soil properties of selected map units. Statements and tables can be similar to those currently used in Florida and elsewhere (Schellentrager et al., 1988) for map unit composition. Table 13-2 represents one possible design. The standard error should be included for users who understand statistics. Users can build confidence intervals at other confidence levels if they so choose. Data acquisition and survey expenses so far exceed publication costs that the survey should contain as much information as possible. The user can decide which information is best suited for the intended application.

Stratification of the transects could show trends in the variability of soil properties across the survey area. These trends could be written into the map

Table 13-2. An example of a possible table for inclusion in a published soil survey.

Map symbol and soil name	Transects	Soil property	Mean†	SE‡	Confidence interval	Confidence level
			cm		cm	
23B2. Leonard silty clay loam, eroded, 2–4% slopes	7	Surface layer thickness	17.1	1.8843	13–21	90%
		Depth to till	35.5	5.4673	28–43	80%
10C2. Armstrong silt loam, 4–9% slopes, eroded. (The Armstrong series is classified as fine, montmorillonitic, mesic Aquollic Hapludalfs.)	6	Surface layer thickness	13.9	2.4855	9–19	90%
		Depth to gray mottles	32.1	4.3245	26–38	80%

† Note that means are listed to one digit beyond significance.
‡ Standard error (SE) is provided to four decimals so that it may be used to calculate confidence intervals at other confidence levels without significant rounding errors.

unit descriptions. It would be confusing to show variances of individual strata in the table, because some significant subdivision of the map unit would be implied. All statistical methodology, including stratification if used, should be clearly defined within the soil survey report.

The confidence interval should not be confused with a prediction interval. The ranges developed here are not prediction intervals for individual observations. Care should be taken not to misrepresent these ranges to the soil survey user.

Expressing a mean and a symmetrical range for a soil property does not completely represent the spatial variability of that property. Specific trends within the delineation are usually apparent to soil survey workers conducting transects. For example, surface layer thickness may be related to slope aspect within delineations in some map units; it may be highly variable over short distances in some map units; or it may gradually decrease towards steeper, downslope margins in other map units. These trends can be briefly discussed in the map unit description and can supplement the soil property range to provide the user with a more realistic, landscape-oriented vision of soil variability.

As soil surveys are digitized to become cartographic databases for use in Geographic Information Systems (GIS), it seems logical to include the data such as Table 13-2 as part of the GIS database. This inclusion could be made at a local level, but no easy mechanism exists for including specific cartographic ranges in the national or statewide databases. Perhaps in the future, ranges of important soil properties can become part of the State Soil Series Database (3SD). Specific ranges could be specified and stored in the Map Unit Use File.

SUMMARY AND CONCLUSIONS

The mean and the variability of soil properties within map units can and should be more clearly, accurately and precisely described to the soil survey user. Because a single mean value implies an assurance that is misleading, confidence intervals for the mean are proposed as the clearest, most accurate way of expressing mean values for important soil properties. Definition and description of ranges for mean values of soil properties should become a standard part of the soil mapping process.

The use of the confidence interval statistic implies the use of data selected in an unbiased fashion. However, it is neither feasible nor desirable to eliminate bias from the soil survey process. We believe that the pragmatic approach described in this paper sufficiently satisfies the underlying assumptions necessary to apply confidence intervals. Statistically knowledgeable soil survey users should be made aware of the bias inherent in this process. Hopefully, future research on data collection within the soil survey framework will reveal how significantly this bias affects the resulting confidence intervals.

The confidence interval is an indication of soil variability; but a need exists to define ranges in soil properties for individual pedons within map units. The tolerance interval provides a useful statistic for defining normally distributed ranges; but appropriate data collection is a major stumbling block in progressive soil surveys. Until a method is developed to efficiently collect data on large numbers of independent, randomly located observations, the use of tolerance intervals in soil survey work is of questionable validity.

Methods and variations other than those described here are possible. Opinions on methodology have been and will continue to be divergent. Exacting consensus seems unlikely and, in fact, may not be a desirable goal, considering the diversity of landscapes and land uses addressed by the NCSS.

Demands on soil surveys are increasing as soil survey users become more sophisticated. The importance of clearly defined, quantified information is increasing with the growing use of digitized, spatial databases. Objectively defined and quantified ranges for the means of soil properties within map units will improve the usefulness of the soil survey for all users. These ranges should be published as an integral part of NCSS soil surveys in the 1990s and beyond.

ACKNOWLEDGMENTS

The authors acknowledge the efforts of Bobby Ward, SCS (presently state soil scientist, Indianapolis, IN), who designed a transect form that efficiently records pertinent transect data and also initiated transecting in the Audrain County, MO soil survey area. We also acknowledge Audrain County Soil Survey party members Alice Geller and Wyn Kelly, who collected the bulk of the transect data which served as a primary inspiration for this paper.

REFERENCES

Ameyan, O. 1986. Surface soil variability of a map unit on Niger River alluvium. Soil Sci. Soc. Am. J. 50:1289-1293.
Arnold, R.W. 1977. CBA-ABC clean brush approach achieves better concepts. p. 61-92. *In* Soil survey quality. Proc. New York Conf. Soil Mapping Quality, Bergamo East, NY. 5-7 Dec. 1977. Cornell Univ. Agric. Exp. Stn. in cooperation with USDA-SCS, Ithaca, NY.
Bascomb, C.L., and M.G. Jarvis. 1976. Variability in three areas of the Denchworth soil map unit. I: Purity of the map unit and property variability within it. J. Soil Sci. 27:420-437.
Bigler, R.J., and K.J. Liudahl. 1984. Estimating map unit composition. Soil Surv. Horiz. 25:21-25.
Bowker, A.H. 1947. Tolerance limits for normal distributions. p. 97-110. *In* C. Eisenhart et al. (ed.) Techniques of statistical analysis. McGraw-Hill Book Co., Inc., New York.
Butler, B.E. 1980. Soil classification for soil survey. Clarendon Press, Oxford, England.
Campbell, J.B. 1978. Spatial variation of sand content and pH within single contiguous delineations of two soil mapping units. Soil Sci. Soc. Am. J. 42:460-464.
Cochran, W.G. 1977. Sampling techniques. 3rd ed. John Wiley & Sons, New York.
de Gruijter, J.J., and B.A. Marsman. 1985. Transect sampling for reliable information on mapping units. p. 150-163. *In* D.R. Nielsen and J. Bouma (ed.) Soil spatial variability. Proc. Workshop ISSS and SSSA, Las Vegas, NV. 30 Nov.-1 Dec. 1984. PUDOC, Waginengan, Netherlands.
Edmonds, W.J., S.S. Iyengar, L.W. Zelazny, M. Lentner, and C.D. Peacock. 1982. Variability in family differentia of soils in a second-order soil survey mapping unit. Soil Sci. Soc. Am. J. 46:88-93.

Edmonds, W.J., J.B. Campbell, and M. Lentner. 1985a. Taxonomic variation within three soil mapping units in Virginia. Soil Sci. Soc. Am. J. 49:394-401.

Edmonds, W.J., J.C. Baker, and T.W. Simpson. 1985b. Variance and scale influences on classifying and interpreting soil map units. Soil Sci. Soc. Am. J. 49:957-961.

Edmonds, W.J., and M. Lentner. 1986. Statistical evaluation of the taxonomic composition of three soil map units in Virginia. Soil Sci. Soc. Am. J. 50:997-1001.

Hole, F.D., and J.B. Campbell. 1985. Soil landscape analysis. Rowman & Allenheld, Totowa, NJ.

Hudson, B.D. 1980. Ranges of characteristics—How valid are they. Soil Surv. Horiz. 21:3,7-11.

Jansen, I.J., and R.W. Arnold. 1976. Defining ranges of soil characteristics. Soil Sci. Soc. Am. J. 40:89-92.

Knox, E.G. 1965. Soil individuals and soil classification. Soil Sci. Soc. Am. Proc. 29:79-84.

Knox, E.G. 1987. Soil series and map units. p. 226. *In* Agronomy abstracts. ASA, Madison, WI.

Lanyon, L.E., and G.F. Hall. 1981. Application of autocorrelation analysis to transect data from a drainage basin in eastern Ohio. Soil Sci. Soc. Am. J. 45:368-373.

McCormack, D.E., and L.P. Wilding. 1969. Variation of soil properties within mapping units of soils with contrasting substrata in northwestern Ohio. Soil Sci. Soc. Am. Proc. 33:587-593.

Mendenhall, W., L. Ott, and R.L. Schaeffer. 1971. Elementary survey sampling. Wadsworth Publ. Co., Inc., Belmont, CA.

Powell, J.C., and M.E. Springer. 1965. Composition and precision of classification of several mapping units of the Appling, Cecil, and Lloyd series in Walton County, Georgia. Soil Sci. Soc. Am. Proc. 29:454-458.

Protz, R., E.W. Presant, and R.W. Arnold. 1968. Establishment of a modal profile and measurement of variability within a soil landform unit. Can. J. Soil Sci. 48:7-19.

Ragg, J.M., and R. Henderson. 1980. A reappraisal of soil mapping in an area of southern Scotland: I. The reliability of four soil mapping units and the morphological variability of their taxa. J. Soil Sci. 31:559-572.

Ransom, M.D., W.W. Phillips, and E.M. Rutledge. 1981. Suitability for septic tank filter fields and taxonomic composition of three soil mapping units in Arkansas. Soil Sci. Soc. Am. J. 45:357-361.

Reybold, W.U., and G.W. Petersen (ed.) 1987. Soil survey techniques. SSSA Spec. Publ. 20. SSSA, Madison, WI.

Schellentrager, G.W., J.A. Doolittle, T.E. Calhoun, and C.A. Wettstein. 1988. Using ground-penetrating radar to update soil survey information. Soil Sci. Soc. Am. J. 52:746-752.

Soil Conservation Service. 1980. Soil survey manual. U.S. Gov. Print. Office, Washington, DC.

Soil Conservation Service. 1983. National soils handbook. U.S. Gov. Print. Office, Washington, DC.

Soil Survey Staff. 1975. Soil taxonomy: A basic system of soil classification for making and interpreting soil surveys. USDA-SCS Agric. Handb. 436. U.S. Gov. Print. Office, Washington, DC.

Steers, C.A., and B.F. Hajek. 1979. Determination of map unit composition by a random selection of transects. Soil Sci. Soc. Am. J. 43:156-160.

Wang, C. 1982. Variability of soil properties in relation to size of map unit delineation. Can. J. Soil Sci. 62:657-662.

Webster, R. 1968. Fundamental objections to the 7th approximation. J. Soil Sci. 19:354-365.

Webster, R. 1977. Quantitative and numerical methods in soil classification and survey. Clarendon Press, Oxford, England.

Webster, R., and P.H.T. Beckett. 1968. Quality and usefulness of soil maps. Nature (London) 219:680-682.

White, E.M. 1966. Validity of the transect method for estimating composition of soil-map areas. Soil Sci. Soc. Am. Proc. 30:129-130.

Wilding, L.P., and L.R. Drees. 1983. Spatial variability and pedology. p. 83-116. *In* L.P. Wilding et al. (ed.) Pedogenesis and soil taxonomy. I. Concepts and interactions. Elsevier, Amsterdam.

Wilding, L.P., R.B. Jones, and G.M. Schafer. 1965. Variation of soil morphological properties within Miami, Celina and Crosby mapping units in west-central Ohio. Soil Sci. Soc. Am. Proc. 29:711-717.

Young, F.J. 1988. Toward the science of soil survey. Soil Surv. Horiz. 29:40-42.

14 Spatial Variability of Organic Matter Content in Selected Massachusetts Map Units

M. Mahinakbarzadeh, S. Simkins, and P. L. M. Veneman

University of Massachusetts
Amherst, Massachusetts

ABSTRACT

The spatial variability of soil organic matter (OM) content as measured by loss-on-ignition, was investigated along several transects and sampling grids in four Massachusetts soil map units. Within four short transects, each located within a single delineation of a soil map unit and sampled at 0.5- or 1.0-m intervals, autocorrelation of the OM content was sufficiently strong to constitute a significant trend that could be removed by curvilinear or nonlinear regression. The regression residuals in these cases evidenced insufficient remaining autocorrelation to merit further analysis using kriging. Organic matter content along a longer transect (1.2 km) located within a single delineation of a map unit sampled at 15-m intervals showed a weaker trend; but the residuals left by regression used to remove trend again showed little autocorrelation. Along a 1.2-km transect crossing the boundaries between three different soil map units, soil OM contents were highly autocorrelated reflecting different OM contents in the soil delineations. These differences formed a pattern too complex to be considered a trend and to be removed by regression. A kriging analysis with jackknifing was able to explain 87% of the variation in OM levels along this transect. Within a two-dimensional grid spanning four soil delineations, OM content was found to follow a positively skewed frequency distribution. The semivariograms of logarithms of soil OM content in different directions within the grid showed evidence of anisotropy, which was removed by dilating the coordinates of the sample locations in one direction. The resulting isotropic semivariogram was used to prepare a map of OM contents within the grid using block kriging. Boundaries on this map resembled map units delineated on the existing 1:15 840 scale soil map.

Reliable estimates of the soil OM content variability require knowledge of its distribution pattern. Estimating the degree of variability and assessing the source of the variance for various soil properties have been the subject of several studies (Warrick & Nielsen, 1980; Hajrasuliha et al., 1980; Gajem et al., 1981; Vieira et al., 1981; Yost et al., 1982; Dahiya et al., 1984; Saddig

---

Copyright © 1991 Soil Science Society of America, 677 S. Segoe Rd., Madison, WI 53711, USA. *Spatial Variabilities of Soils and Landforms.* SSSA Special Publication no. 28.

et al., 1985; Webster, 1985; Riha et al., 1986; Trangmar et al., 1986, 1987; Schimel et al., 1988; West et al., 1989). Most of these efforts considered a physical, chemical, or geomorphological aspect of the landscape. In general, parent material, climate, topography, various physical and chemical processes, and biological activity are known sources of variability in the natural landscape.

Considerable effort has been extended to estimate the variability of soils with a limited number of observations under the assumption that measured properties represent the variability of unsampled locations in the landscape. For example, soil-survey characterization studies typically are conducted by collecting point samples and assuming that they represent neighboring points and that the observations are random and spatially independent regardless of their location in the field. With this approach, mean, variance, standard deviation, and coefficient of variation for various soil properties can be estimated. This method, however, does not necessarily provide a picture of the true variation of the soil properties because the calculated variance does not consider the distance between observations (Saddig et al., 1985). In addition, a normal distribution often is assumed that may not be truly representative.

In recent years, a new approach to spatial analysis has been developed and commonly referred to as the geostatistical method (Matheron, 1963). This technique, also known as regionalized variable theory, considers differences between soil properties at different locations (Webster, 1985). Geostatistical methods provide a means to understand the spatial dependency of soil variability that cannot be obtained from standard frequency distribution analysis.

The degree of heterogeneity of soil in a landscape determines the spatial dependency of any soil property. In general, the degree of dependency for any soil property varies within a wide spectrum of which randomness represents one end, and the presence of strong trends represents the other end. Within these two extremes, the values of parameters may not be independent of each other and application of spatial analysis then is the appropriate framework for expressing these relationships.

Knowledge of the variability in soil OM content is essential in predicting the soil chemical behavior in regard to the behavior of fertilizers, pesticides, and other organic chemicals. This paper reports the spatial variability of soil OM content assessed in selected soil map units representing different parent materials in Massachusetts.

MATERIALS AND METHODS

Two delineations of each of the following soil map units were sampled in Hampshire County: Agawam fine sandy loam, 0 to 3% slope; Hadley silt loam; and Limerick silt loam. Hampshire County is located in central

Massachusetts and was mapped according to modern survey standards at a scale of 1:15 840 (Swensen, 1981). In addition, two delineations of Charlton fine sandy loam, 8 to 20% slopes were sampled in Franklin County just north of Hampshire County. Franklin County was surveyed during the period 1947 to 1963 (Mott & Fuller, 1967).

The *Agawam* series (Typic Dystrochrepts, coarse loamy over sandy, mixed, mesic) are developed in well-drained glaciated outwash; *Hadley* series (Typic Udifluvents, coarse silty, mixed, nonacid, mesic) occur in well-drained alluvial deposits; the *Limerick* series (Typic Fluvaquents, coarse-silty, mixed, nonacid, mesic) are formed in poorly or somewhat poorly drained alluvial deposits; and *Charlton* series (Typic Dystrochrepts, coarse loamy, mixed, mesic) are found on uplands in well-drained glacial till. Delineation size ranged from 5.5 ha for one of the *Charlton* units to 32.5 ha for one of the *Hadley* delineations.

All delineations have been under cultivation for decades mostly in sweet corn (*Zea mays* L.). Mean annual air temperature of the region is 8.9°C (Swensen, 1981). Mean annual precipitation is 115 cm evenly divided throughout the year, with approximately 40% of precipitation occuring during the growing season. The delineations were sampled at various intervals along transects or in grid patterns. At each soil sampling point, 2.5-cm-diam. cores were obtained from the upper 0 to 15 cm of the A horizon. Soil OM was evaluated by the loss-on-ignition method (Davies, 1974).

Different sampling intervals to assess spatial distribution of OM content within delineations were evaluated by sampling the delineations of the *Charlton* and *Hadley* map units along one 50-m transect at 0.5-m intervals and one 80-m transect at 1-m intervals. These short transects within a delineation of each map unit were situated approximately perpendicular to each other in north-south and east-west directions. To evaluate possible autocorrelation along a long transect, a delineation of the *Hadley* silt loam map unit was sampled at 15-m sampling intervals in two 1200-m-long transects perpendicular to each other.

Duplicate delineations of Hadley silt loam, Limerick silt loam, and Agawam fine sandy loam map units were sampled in grids at 15-m intervals to assess the within-delineation variability. Location of the grids within each delineation was selected at random, and total number of samples within each grid ranged from 60 to 100.

The spatial variability in OM distribution along a transect crossing different delineations was determined along a 1200-m transect sampled at 15-m intervals across Hadley, Limerick, and Agawam delineations. To evaluate the spatial dependency of soil OM content in two dimensions, a cultivated plot of land containing delineations of Hadley silt loam and Limerick silt loam map units was sampled at 15-m intervals in a grid pattern with 8 rows and 40 columns.

RESULTS AND DISCUSSION

Variation of Soil Organic Matter Content in Short Transects

Distribution of OM content along the short transects (50 and 80 m) within single delineations of map units at 0.5- and 1-m sampling intervals showed strong trends. Semivariance analysis of the residuals left when the trend was removed by regression indicated periodicity within transects. The semivariogram for the Hadley transect (Fig. 14-1) showed that measured values for OM content taken about 6, 24, and 32 m apart showed greater dissimilarity than measurements taken about 15 and 26 m apart. Samples at 40-m intervals tended to show greater similarity. The semivariogram in one Charlton transect indicated that locations about 1, 32, 43, and 49 m apart were more similar than those 15 and 45 m apart; whereas, in a second Charlton transect values 10, 20, 28, 40, and 47 m apart tended to be more similar than values 5, 15, and 35 m apart. In all semivariograms, the ratio of sill values to nugget values was low. The ratio was about 0.5/0.4 for the Hadley transect, 0.2/0.1 for the first Charlton transect, and 0.21/0.15 for the second Charlton transect. These low values indicated little autocorrelation among the measured values.

The general structure of the semivariograms calculated before trend removal showed that within single delineations of the map units studied, the distribution of OM content exhibited a strong spatial dependency. Application of standard statistical analysis procedures such as regression then is more appropriate for variability analysis. Based on the observation of existing trends in the short transects, suitable sampling intervals for the longer transects were determined to be 15 m. By selectively removing samples from the

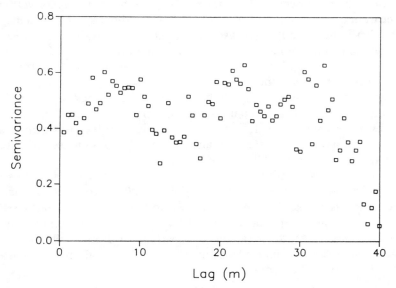

Fig. 14-1. Semivariogram of residual OM content along a transect in a Hadley map unit based on 1-m sampling intervals.

data base, i.e., selectively varying the distance between sampling points, autocorrelation was observed. Relative to the requirement of at least 80 samples for geostatistical analysis, the minimum length of a transect crossing several delineations was estimated to be 1200 m.

Variation of Soil Organic Matter Content in Long Transects

The soil OM content distribution along the two 1200-m transects within a delineation of the Hadley silt loam map unit exhibited a significant trend but little autocorrelation along both transects (Fig. 14-2). The semivariogram (Fig. 14-3) for the east-west direction still revealed periodicity. The nugget value was almost one-half of the maximum of the semivariance. The ratio of sill over nugget value was about 0.14/0.08. Even though the calculated semivariogram for the north-south direction showed less sill/nugget effect in comparison to the east-west direction; the low ratio of sill over nugget value, 0.06/0.02, was not large enough to justify geostatistical methods to estimate OM distribution, regardless of the sampling intervals. It was concluded that in the single delineation of the map unit studied, classical statistical methods were more appropriate to evaluate variability in soil OM content.

Variation of Soil Organic Matter Content Within and Between Delineations

The variability of soil OM content within and between individual delineations of Hadley silt loam, Limerick silt loam, and Agawam fine sandy loam map units was measured with classical statistics such as analysis of var-

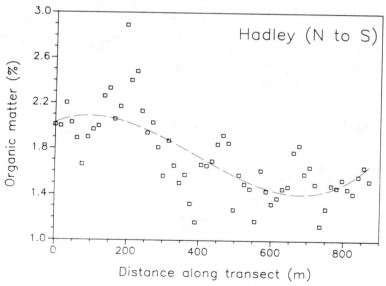

Fig. 14-2. Evidence of trend in the distribution of OM content along a north-south transect in a Hadley map unit at 15-m sampling intervals.

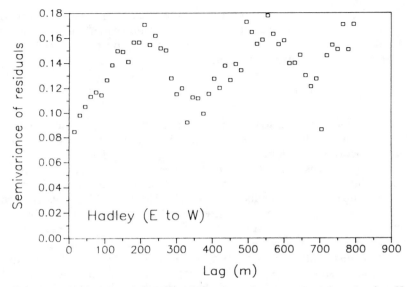

Fig. 14-3. Semivariogram calculated for OM content along an east-west transect in a Hadley map unit based on 15-m sampling intervals.

iance, the coefficient of variation (CV), and the t-test. There was no significant difference between the two Hadley delineations but differences between the Agawam and Limerick map units were significant (Table 14-1).

By selecting samples based upon sampling intervals at 30, 45, and 60 m in each of the studied grids, the same statistical methods were employed. By increasing the sampling interval, no significant differences were observed within the Hadley and Limerick delineations with respect to OM variability. A low t-ratio in the Hadley delineation indicated a small degree of variability in OM content, implying that relatively small numbers of samples are sufficient to reasonably evaluate the variability in OM content.

The variability in OM content between delineations of various map units indicated that delineations of the Hadley map unit, in comparision to Agawam delineations, were not statistically different. In contrast, there were

Table 14-1. Summary of the statistical analysis of OM content within map units based on grid sampling at 15-m intervals.

Map unit	Mean	SD	CV%	t ratio	DF
Hadley 1	4.54	1.09	23.94	1.64	92
Hadley 2	4.92	1.62	32.88		
Agawam 1	3.43	0.38	11.07	19.92**	111
Agawam 2	6.10	1.53	25.04		
Limerick 1	7.50	5.44	71.48	4.14**	151
Limerick 2	4.96	2.86	57.66		

** Significant at the 0.01 probability level.

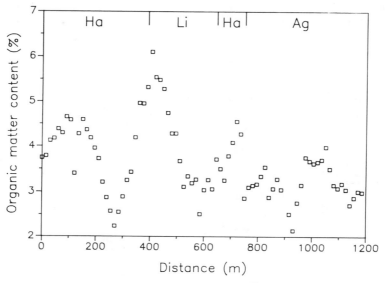

Fig. 14-4. Distribution of OM content along a transect including Hadley (Ha), Limerick (Li), and Agawam (Ag) map units; sampling intervals were 15 m.

significant differences between the Hadley vs. Limerick, and Agawam vs. Limerick delineations.

Soil Organic Matter Content Variation in Multidelineation Transects

In transects spanning more than one delineation, variability in OM content usually cannot be evaluated in the same fashion as single delineations of map units. Differences in OM content were assessed at 15-m intervals along a 1200-m transect crossing several delineations including Hadley, Limerick, and Agawam map units (Fig. 14-4). The general pattern along the transect indicated less trend, hence greater autocorrelation. Geostatistics then were employed to determine the spatial dependency of the soil OM content.

Experimental semivariograms were calculated (Robertson, 1987) for lag distances of 15 m using the number of data pairs up to lag 600 m. The semivariogram showed evidence of periodicity. A high ratio of sill to nugget value (0.7/0.5) also was evident indicating appreciable autocorrelation within the transect. The semivariogram exhibited very little nugget effect showing that variation due to sampling error or the presence of short-range spatial dependency is smaller than the variation that is spatially dependent over a range greater than the sampling intervals.

Jackknifing is a procedure in kriging in which values at a measured site are estimated from values from surrounding known sites, and the estimated values are compared to the measured data (Vieira et al., 1983). This technique was used to predict values for OM content at sampled locations within the

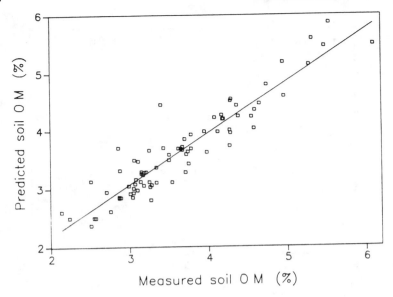

Fig. 14–5. Regression relationships between measured soil OM and predicted soil OM content using kriging within the multidelineation transect, including Hadley, Limerick, and Agawam map units.

transect. Fitted plots of measured and predicted values showed a good correlation (Fig. 14–5).

Organic matter content variability in two dimensions was assessed in the 108- by 585-m^2 grid sampled at 15-m intervals, and crossing delineations of Hadley silt loam and Limerick silt loam map units. The three-dimensional representation of OM content vs. spatial location is shown in Fig. 14–6. Grid ends represent Hadley delineations, and the middle section with higher OM content values, corresponds to the delineation of the somewhat poorly drained Limerick map unit.

One of the basic requirements for kriging is a normal distribution of the population. The cumulative frequency of measured OM content revealed that the actual data did not completely fit a normal distribution curve. Logarithmic transformations of the percentage OM yielded a better normalized data set for use in subsequent statistical analysis. Semivariograms based on the transformed OM values had a shorter range (Fig. 14–7) in the x-direction rather than in the y-direction that indicated anisotropy. To fit the semivariograms for kriging applications, the lag distance in the x-direction was expanded by a factor of 2.75. An exponential model (Webster, 1985) was fitted to the pooled semivariogram. Modeling of the semivariograms permitted identification of the changes in spatial variability with respect to the direction in the study site, which in turn, is a reflection of differences in drainage class and landscape position. A significant ($P < 0.001$), although not very strong correlation ($r^2 = 0.438$) between jackknifed and measured values for OM content was found in the grid.

SPATIAL VARIABILITY OF ORGANIC MATTER IN MASSACHUSETTS 239

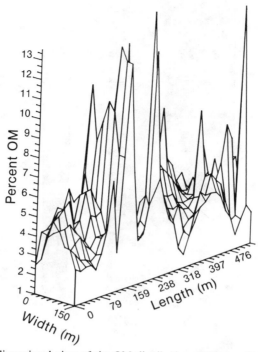

Fig. 14-6. Three-dimensional view of the OM distribution within a grid including Hadley-Limerick-Hadley map units.

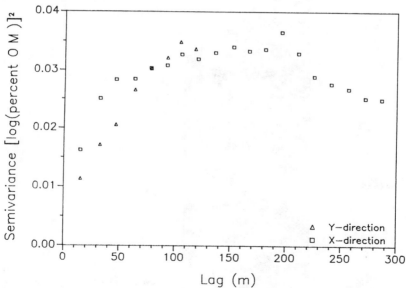

Fig. 14-7. Evidence of anisotropy of logarithmically transformed OM values in the Hadley-Limerick-Hadley grid.

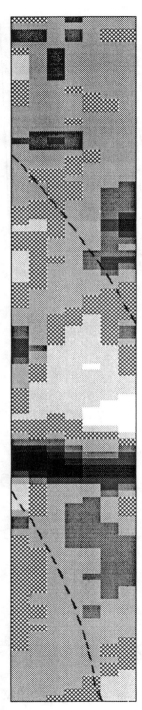
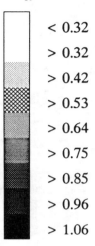

Fig. 14–8. Map of Hadley grid showing average logarithmically transformed OM levels calculated by block kriging.

Logarithmically transformed values of OM content were block kriged in 5.45- by 15-m^2 cells over the studied grid (Fig. 14–8). The range of kriged values is shown as a spectrum of shaded colors. Dark regions across the grid indicate high OM content concentrations, and lighter colors show sites of lower concentrations. The pattern of kriged values demonstrates the heterogeneous distribution of OM in the landscape. The highest values mostly coincide with the somewhat poorly drained Limerick delineation, whereas the other regions are representative of delineations of the well-drained Hadley silt loam map units.

The spatial OM distribution pattern and its heterogeneity can be explained by differences in map unit composition. In general, in transects within a single delineation of a map unit, strong trends were evident, precluding simple kriging. The variance in OM content in the Hadley and Agawam delineations was small. In contrast, the variability in the delineations of the Limerick map unit was quite large. Using semivariograms, analysis of the spatial dependence of OM content in short and long transects in single delineations indicated that the variation in soil OM content showed trends. Sampling intervals at 0.5, 1, and 15 m along both short (40- and 80-m) and long (1200-m) transects indicated that geostatistics should not be applied in delineations of map units with a relatively uniform composition. On the other hand, classical statistics are less applicable in delineations of map units comprised of soils with contrasting soil characteristics such as in complexes or associations.

Both the transect and grid crossing delineations of several map units showed a more pronounced spatial dependency for OM content. Organic matter did not accumulate to the same degree in all delineations studied perhaps because of differences in drainage class. Hadley and Agawam map units represent deep, well-drained soils. The lack of strong redoximorphic features in the subsoil indicated good aeration, which may have accelerated OM decomposition. On the other hand, the Limerick delineations were comprised of deep, poorly, or somewhat poorly drained soils on floodplains. The presence of redoximorphic features in the subsoil showed seasonal reducing conditions that resulted in lower mineralization rates in the soil OM.

CONCLUSIONS

The distribution of OM content along short and long transects, and also within and between single delineations of map units, was found to exhibit strong trends, regardless whether 0.5-, 1-, or 15-m sampling intervals were used. Such a condition may be statistically analyzed by taking random or grid samples and applying classical statistics to provide mean, standard deviation, variance, standard error, regression, and coefficient of variation as estimators in the variability of OM content. Conversely, at locations where delineations of several map units or quite variable soils are considered, they are likely to have significant autocorrelation in the OM distribution. In our study, geostatistical analysis using 15-m sampling intervals and kriging to predict values for unsampled locations was satisfactory for the evaluation of spatial dependency of OM content.

REFERENCES

Dahiya, I.S., J. Richter, and R.S. Malk. 1984. Spatial variability: A review. Int. J. Trop. Agric. 11:1–102.

Davies, E.B. 1974. Loss-on-ignition as an estimate of soil organic matter. Soil Sci. Soc. Am. Proc. 38:150–151.

Gajem, Y.M., A.W. Warrick, and D.E. Myres. 1981. Spatial dependence of physical properties of a Typic Torrifluvent soil. Soil Sci. Soc. Am. J. 45:709–715.

Hajrasuliha, S., N. Baniabbasi, J. Metthey, and D.R. Nielsen. 1980. Spatial variability of soil sampling for salinity studies in south-west Iran. Irrig. Sci. 1:197–208.

Matheron, G. 1963. Principles of geostatistics. Econ. Geol. 58:1246–1266.

Mott, J.R., and D.C. Fuller. 1967. Soil survey of Franklin County, Massachusetts. USDA-SCS, Washington, DC.

Riha, S.J., B.R. James, G.P. Senesca, and E. Pallant. 1986. Spatial variability of soil pH and organic matter in forest plantations. Soil Sci. Soc. Am. J. 50:1347–1352.

Robertson, G.P. 1987. Geostatistics in ecology: Interpolating with known variance. Ecology 68:744–748.

Saddig, M.H., P.J. Wierenga, M.H. Hendrickx, and M.Y. Hussain. 1985. Spatial variability of soil water tension in an irrigated soil. J. Soil Sci. 140:126–132.

Schimel, D.S., S. Simkins, T. Rosswall, A.R. Mosier, and W.J. Parton. 1988. Scale and measurement of nitrogen-gas fluxes from terrestrial ecosystems. p. 179–193. *In* T. Roswall et al. (ed.) Scale and global changes. John Wiley & Sons, New York.

Swensen, E. 1981. Soil survey of Hampshire County, Massachusetts, central part. USDA-SCS, Washington, DC.

Trangmar, B.B., R.S. Yost, and G. Uehara. 1986. Spatial dependence and interpolation of soil properties in West Sumatra, Indonesia: I. Anisotropic variation. Soil Sci. Soc. Am. J. 50:1391–1395.

Trangmar, B.B., R.S. Yost, M.K. Wade, G. Uehara, and M. Sudjadi. 1987. Spatial variation of soil properties and rice yield on recently cleared land. Soil Sci. Soc. Am. J. 51:668–674.

Vieira, S.R., D.R. Nielsen, and J.W. Biggar. 1981. Spatial variability of field-measured infiltration rate. Soil Sci. Soc. Am. J. 45:1040–1048.

Vieira, S.R., J.L. Hatfield, D.R. Nielsen, and J.W. Biggar. 1983. Geostatistical theory and application to variability of some agronomical properties. Hilgardia 51:1–75.

Warrick, A.W., and D.R. Nielsen. 1980. Spatial variability of soil physical properties in the field. p. 319–344. *In* D. Hillel (ed.) Application of soil physics. Academic Press, New York.

Webster, R. 1985. Quantitative spatial analysis of soil in the field. Adv. Soil Sci. 3:1–70.

West, C.P., A.P. Mallarino, W.F. Wedlin, and D.B. Marx. 1989. Spatial variability of soil chemical properties in grazed pastures. Soil Sci. Soc. Am. J. 53:784–789.

Yost, R.S., G. Uehara, and R.L. Fox. 1982. Geostatistical analysis of soil chemical properties of large land area. I. Semivariogram. Soil Sci. Soc. Am. J. 46:1028–1032.

15 Geographic Information Systems for Soil Survey and Land-Use Planning

R. David Hammer, Joseph H. Astroth, Jr., and G. S. Henderson

University of Missouri
Columbia, Missouri

Fred J. Young

USDA-SCS
Columbia, Missouri

ABSTRACT

Land-use planners are requiring site-specific information too detailed for inclusion in published soil surveys. Special-use maps are receiving increasing attention as management alternatives to the soil survey. The special-use maps are often generated with Geographic Information Systems (GIS). Soil survey information frequently is digitized as a part of the data base for special-use maps. Slope and aspect information commonly is generated from Digital Elevation Models (DEM). Field verification of digitized soil–landscape maps is lacking. The opportunity exists to create confusion and ambiguity by layering databases of unknown accuracy and precision. Ground-truth measurements should be obtained to verify precision of computer-generated special–use maps. Soil series, seasonal distribution of soil water, and plant growth and yield are highly correlated to geomorphic surfaces. Digital Elevation Models and GIS have great potential as research tools to enhance our understanding of the process involved in evolution of soil landscapes.

Recent years have witnessed a proliferation of applications of GIS by federal, state, and municipal government agencies for natural-resource planning, land-use planning, and environmental protection. Soil surveys are a frequent component of such GIS applications. Soil-based applications of GIS technology introduce new demands upon soil surveys and the information they contain, and produce new users of soil survey information. Many users of GIS have limited knowledge of soil science and little previous experience

---

Copyright © 1991 Soil Science Society of America, 677 S. Segoe Rd., Madison, WI 53711, USA. *Spatial Variabilities of Soils and Landforms.* SSSA Special Publication no. 28.

with soil surveys. These users may be unaware of either the potentials or limitations of soil survey information.

Two examples of misunderstanding are perceptions of map unit purity and of data in interpretive tables. New users of soil surveys often are unaware that delineations on soil maps are not pure. Limiting and nonlimiting inclusions and the range of characteristics of a soil series may not be considered by anyone incorporating digitized soil data into a GIS file. The uninitiated may assume that the modal pedon described in the soil survey represents the mean or modal attributes of most of the soil properties in the map unit. Interpretive tables in published soil surveys can contribute to user misunderstanding. The sources of data used in interpretive tables frequently are not supplied. Yield predictions and other tabular data are not accompanied by error terms (standard deviations or confidence intervals). The problems of educating soil survey users can be exacerbated when data are provided in digitized data sets.

Arnold (1987) has cautioned that " . . . the kinds of map units, the level of detail of soil classification in the legend, and the detail in which map units are described in the report have as much impact on the usefulness of a soil survey as the detail on the soil map." His concern seems particularly apropos in the realm of GIS and land-use planning. Conversely, GIS holds much promise as a technology to enhance pedological research and to improve the quality and precision of the soil survey effort.

OBJECTIVE

This chapter will address the problems and potentials of GIS as a set of tools for soil science research. We will define GIS, inform readers of additional sources of GIS information, and briefly review some current applications of GIS and soil survey for land-use planning. Most importantly, we will focus upon the unique potential of GIS to enhance the quality of soil survey. Emphasis in enhancing the quality of soil survey will be placed upon forest land and site productivity, which is intimately related to topography, and to assessing soil–landscape relationships. The latter of these two topics has tremendous potential for pedological research.

An enormous volume of literature that generally has been neglected addresses deciduous forest site productivity. The application of GIS to forest-site quality in soil survey is but one example of how a new technology can be used to apply previously acquired information to solve problems.

The goal of the manuscript is to prompt dialogue, with the hope that the science of pedology and the making of soil surveys can benefit from wise application of GIS technology. A companion paper (Young et al., 1991) suggests methods to increase precision of map unit description while building a soil-spatial database that would be suitable for incorporation into a GIS system.

GEOGRAPHIC INFORMATION SYSTEMS DEFINED

The rush by a variety of users to apply GIS technology to discipline-specific applications has resulted in some confusion regarding the definition of GIS. Parker (1987) and Cowen (1988) have provided excellent reviews of both the situation and the relevant literature. They included discussions of the several definitions of GIS. Parker (1987) attributed some of the confusion regarding nomenclature and definition to the variety of applications of GIS technology. He pointed out that resource management, the arena most ardently embracing GIS applications, is as much an art as a science. Resource management itself " . . . not a well-defined process . . . is largely dependent on local ecology, background of the resource manager(s) involved, management history, and management objectives, all of which vary spatially and temporally." Parker considered GIS to be a *technology* with important and

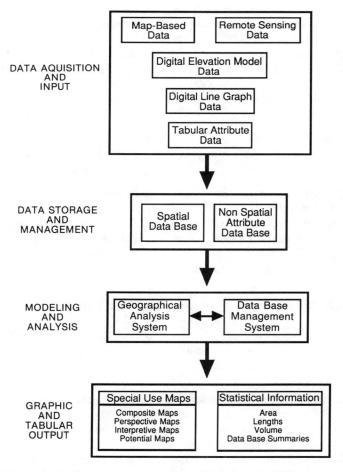

Fig. 15-1. Schematic drawing of the components and function of a GIS.

specific applications for spatial data processing. Cowen (1988) distinguished the differences between computer mapping and the spatial analysis attributes of GIS.

New maps produced by overlaying several existing map layers into a GIS effectively constitute new information. The relationships revealed through GIS were previously unknown or not so clearly defined. This computer cartography application has received considerable attention because of its relative ease of use. Soil surveys commonly have been digitized and incorporated into layered databases. Tomlinson (1972) stated that GIS is not a unique field of endeavor, but is the "common ground" between information processing and those with a need for spatial analysis.

Burrough's recent (1986) monograph contains a definition of GIS that adequately encompasses the variety of uses to which GIS is being subjected. Burrough considered that a complete GIS is a "set of tools for collecting, storing, retrieving at will, transforming, and displaying spatial data from the real world for a particular set of purposes." According to Burrough, a GIS contains three components—the computer hardware, application software, and an organizational context. Included in his description of "organizational context" was " . . . the retraining of personnel and managers to use the new technology in the proper . . . context" (Burrough, 1986). The inclusion of "organizational context" is the key dimension separating this definition of GIS from Parker's definition (1987). Figure 15-1 is a schematic of the components of GIS without the organizational context.

ADVANTAGES OF GEOGRAPHIC INFORMATION SYSTEMS FOR LAND-USE PLANNING

Commercially available software provides the GIS user with three different, but related, fundamental applications: database management, computer cartography, and spatial analysis. These applications will be briefly discussed in a general sense. Their applications to soil survey, land-use planning, and pedological research will be developed more completely within the context of application examples.

Geographic Information Systems Computer Cartography Attributes

The computer cartography capabilities of GIS (Rhind, 1977, as cited by Burrough, 1986) include the abilities to:

1. Make existing maps more quickly and at lower cost.
2. Make user-specific maps.
3. Make maps in the absence of skilled cartographers.
4. Allow rapid evaluation of various methods of displaying data.
5. Rapidly incorporate new digital data into the system to upgrade existing maps.
6. Enhance analysis of data requiring interaction between statistical analysis and mapping.

7. Minimize the effects of classification and generalization on the quality of the stored data.
8. Produce maps difficult to create by hand, such as three-dimensional surface maps.
9. Enhance consistency of selection and generalization procedures.
10. Effect cost savings and quality improvement through automation.

Spatial Analysis with Geographic Information Systems

We are of the opinion that spatial analysis attributes of GIS potentially offer great promise to expand the quantitative databases of pedology and geomorphology. More importantly, the potential exists to weld together these related sciences, through three-dimensional analysis and displays, in ways not previously possible. Zakrjewska's (1967) review of landform analysis provides an excellent template for possible GIS applications in soil and landform relationships.

Potential of Geographic Information Systems for Land-Use Planning

Spatial analysis with unique capabilities for cartographic database layering is a unique GIS technology that offers an ideal method to incorporate soil survey data into management plans. The manager or scientist can use GIS to overlay existing databases and to transform the data and/or investigate spatial relationships among the various data. Published soil surveys have been used with GIS technology for a variety of post-survey soil interpretation and land-use applications.

Walsh (1985) was among the first to discuss the potential of GIS with soil, geologic, and topographic data for land-use planning. He recognized the virtually unlimited potential of GIS in this context. From Walsh's perspective, the chief advantages of GIS are the ability to:

1. Correlate vegetation and topographic information with environmentally important attributes such as drainage basin acreage and terrain configuration.
2. Integrate diverse cartographic databases such as conventional maps and Landsat imagery.
3. Quickly and creatively combine and retrieve stored data in various combinations.

Walsh also commented on the potential of GIS for data collection and monitoring " . . . over extensive areas and on a repetitive schedule." Analytical modeling and trend analysis are particularly important uses of this potential application.

REMOTE SENSING AND IMAGERY ANALYSIS

Remote sensing and imagery analysis have received considerable attention in the literature. Our purpose is not to review this extensive body of research, but to highlight specific applications that have shown promise for soil science.

Thematic Mapper (TM) data have been combined with 30 m digital terrain data in Wisconsin (Lee et al., 1988). The data were used to identify soil boundaries in hilly terrain. Results showed that general classification of soils is possible. However, accuracy did not equal that obtained with traditional soil survey techniques. The data were most effective when combined through additive rather than layered techniques.

Although percentage accuracy declined with increasing canopy cover of corn (*Zea mays* L.) and soybean [*Glycine max* (L.) Merr.], TM data was useful in classifying 60% of soils analyzed in Iowa to the proper subgroup level. Applications appeared to be most useful for regional mapping (Thompson & Henderson, 1984).

Combinations of soil and vegetation parameters accounted for much of the variance in satellite imagery spectral bands in rangeland mapping. Elevation, surface stoniness, soil surface pH and clay content, aspect, and litter cover all were important in separating Arizona rangeland soils with satellite imagery (Horvath et al., 1984). Stoner and Baumgardner (1981) concluded that characteristic variations in reflectance of surface soil properties related to wetness, soil-moisture regimes, parent material, and vegetation were useful in separating soils at the higher categories of soil taxonomy. Su et al. (1989) demonstrated that DEM data could be used with Landsat TM data to benefit second-order soil surveys of rangeland soils.

Many of the soil-vegetation parameters shown to be important for separating soils with remotely sensed data are attributes related to soil-landscape features such as topographic position, landform surface shape, elevation, and aspect. As will be discussed later, these parameters can be identified with the aid of GIS technology and DEM data.

INCORPORATIONG DIGITAL ELEVATION MODELS INTO A GEOGRAPHIC INFORMATION SYSTEM FOR SOIL RESEARCH

"Any digital representation of the continuous variation of relief over space is known as a digital elevation model" (Burrough, 1986). A digital terrain model (DTM) differs from a DEM by containing landscape features in addition to elevations of land surface points.

Burrough (1986) presented a list of 10 GIS uses for DEM. The most important DEM applications for soil survey and land-use planning include:

1. Three-dimensional display of landforms.
2. Planning roads.
3. Provide image simulation models of landscape processes.
4. Use as a template upon which to overlay single-use data such as septic field suitability, vegetation distribution, or land values.
5. Statistical comparisons of terrain features.
6. Use in developing slope and aspect maps.
7. Delineating specific land surface segments for analyses—identifying watersheds and their drainage networks, for example.

Unfortunately, DEM's are not widely available. The U.S. Geological Survey (USGS) is producing 7.5-min DEM's on a 30-m base for the entire USA, but these are currently available only for selected quadrangles (Elassal & Caruso, 1984). The USGS DEM's generally have been produced from manually or automated scanning of photographs taken from an average altitude of 12 200 m. Elassal and Caruso (1984) reported that vertical accuracy of 7.5-min DEM's produced from aerial photography can be expected to range from 7- to 15-m standard deviation. Accuracy of individual quadrangles is dependent upon the number of survey control points obtained, resolution of the source photographs, and image processing methodology.

Efforts are underway to develop DEM's from images taken by the French SPOT (Systeme Probatoire d'Observation de le Terre) satellite. Specific DEM developmental methodology from SPOT imagery has been reported (Swann et al., 1987). Results from preliminary efforts indicate that automated DEM's could be produced from SPOT imagery with vertical accuracy of 7-m standard deviation. Investigators reported that vertical accuracy was improved to 4.6-m standard deviation with low radiometric variation and increased terrain relief (Swann et al., 1988).

APPLICATIONS OF GEOGRAPHIC INFORMATION SYSTEMS WITH SOIL SURVEY

Applications of GIS technology in soil survey indirectly impact land-use and natural-resource planning. Soil survey is often a layer in GIS applications, which range from land-use and natural-resource planning at local municipality levels to global resource inventory. Only selected examples from this body of literature are reviewed within this chapter. The reader is advised that many published reports of GIS applications are in nonrefereed symposia proceedings. Often these reports discuss applications of projects without detailing the results. Many reported projects are relatively recent, and are planned to be long-term, but are not yet completed. Appendix A lists several sources of reported GIS applications.

Utilizing soil survey data in a GIS requires building the digitized soils database (DSD), incorporating the DSD into a comprehensive GIS base, then applying the DSD for the stated objective. This activity is underway in several states.

In North Carolina, databases containing soil information are being incorporated into GIS as well as being published as traditional soil surveys. Planned uses of the digitized soil surveys include: (i) a soil-based system for land taxation, (ii) a component of a long-term water quality research effort, (iii) development of special-use soil interpretive maps, and (iv) coastal planning and management (Hermann & Floyd, 1986). The importance of soil-landscape and soil-land-use data in GIS format should be especially important in urban planning. Special urban planning applications would exist after a catastrophic natural disaster such as Hurricane Hugo.

Logan et al. (1982) used GIS with soil, land-use, and topographical information to estimate potential soil erosion from the Lake Erie basin in

Ohio under a variety of possible tillage and crop management systems. The Universal Soil Loss Equation was used with the information from the layered database to produce the potential soil erosion figures. The GIS technology allowed a much more rapid and precise evaluation of land-use options than would have been possible with conventional means.

British Columbia is building a province-wide database including information on climate, surficial geology, soils, and aquatic systems that will be layered over a DTM. Planned applications of the British Columbia project include the prediction of waterway siltation as a result of logging practices and monitoring acid rain effects on soils (Balser & Fila, 1987).

Susceptibility to erosion of rangeland converted to cultivation was assessed with GIS in South Dakota (Best & Westin, 1984). A grid-based GIS was used to manipulate a database composed of land use interpreted from Landsat imagery, soil capability class, and slope. Wind-erodibility group ratings and vegetation-suitability ratings were also included and based on interpretive tables for established soil series. The GIS technology allowed a rapid, cost-effective identification of soil/landscape combinations unsuitable for cultivation. Special-use maps for management planning were also produced with GIS (Best & Westin, 1984).

Other applications of GIS with soil information or remotely sensed data include urban planning (Fitzgerald, 1987), water quality monitoring (Whitehead, 1986), wildlife habitat planning and inventory (Murray & Leckenby, 1985), rangeland planning (Rush et al., 1985), planning and monitoring agricultural production (Smith et al., 1987), timber inventory (Mead & Rowland, 1987), monitoring insect infestation (Parks et al., 1987; Ahern et al., 1985), and route planning for logging access (Berry, 1986). Although results have not been reported, GIS is being used to aid assessment of acid rain impacts on water resources and forest site quality.

Geographic Information Systems in the Improvement of Soil Survey

No studies have reported GIS user requirements for accuracy and precision of soil survey data. The possibility exists that high intensity GIS land-use applications, such as urban planning, will require from soil surveys a degree of precision that was not incorporated into map unit design. Even in the unlikely absence of such future requirements, GIS technology may offer soil scientists an exciting opportunity to improve the quality of their surveys while conducting them in a more cost-effective manner.

Current Field Soil Survey Techniques

Soil survey field mapping techniques have changed little during the past 25 to 30 yr. Although technological advances such as ground-penetrating radar, Landsat imagery, other remote-sensing methods, and GIS have been introduced to the profession; the field soil scientist continues to traverse the landscape with auger or probe and aerial photographs. The field soil scientist makes the important decisions for the soil survey. The field soil scientist also determines map unit composition and boundaries, determines slope ranges and breaks, and decides the location and frequency of soil-sampling locations.

The conscientious surveyor will utilize as much relevant information about the survey area as can be found, including: topographic maps, geologic surveys of various kinds, previously published surveys from the survey area or areas with similar parent materials, climate and geomorphology, published research conducted on the soils of the survey area or upon similar soils, maps of native vegetation, laboratory analyses of soil samples, and climatic data. The available maps may be of different scales or may lack cultural features. The different maps may be difficult to compare with one another, and precise identification of locations within the survey area may be difficult.

The conventional resources can be incorporated, compared, and integrated with GIS technology. Most existing data is on paper maps, which must be digitized and edited prior to incorporation into GIS. In many cases, these data must be recompiled in a rectified basis.

Examples include: "premaps" of slope, aspect, and elevation developed from DEM's, premaps combining landform morphology and geologic strata, premaps of precipitation and temperature in mountainous terrain, or premaps of land use and geomorphology. Soil properties important for land use could be combined from databases for established soil series and used in conjunction with DEM's within GIS to produce special feature maps relating soil properties (such as surface texture or depth to fragipan) with geomorphic surfaces. Subsequent to being incorporated into a GIS, these resources remain as permanent, flexible, computer files.

Figure 15-2 is a three-dimensional surface contour of the Missouri River floodplain and adjacent loess bluffs in Atchison County in northwest Missouri. The data used to produce this image were 10-m horizontal interval DEM's manually developed from stereo photography taken at an altitude of 1220 m. The parallel lines in the floodplain west of the loess bluffs are the roadbed of Interstate 29. Vertical relief in this landscape was approximately 32 m from the floodplain to the highest summit in the loess hills. The vertical precision of these DEM data are being investigated. The research will also assess the accuracy of slope and aspect maps produced with the DEM and ARC/INFO software. The potential uses to the soil surveyor of a three-dimensional surface contour map should be obvious. Figure 15-2 could be enhanced by identifying and overlaying slope classes, vegetative cover, potential land use, or other information.

The potential uses of GIS and DEM as aids to soil survey have received little attention. The presurvey incorporation of existing topological, vegetative, geologic, and climatic data into a GIS system to produce special-use maps for the soil survey crew is a possible attribute to soil survey. The utility of this approach may be enhanced in regions or areas where the completed soil survey will likely be incorporated with these thematic maps into a GIS for post survey use. This approach may also be useful where terrain restricts access to the survey area.

Klingebiel et al. (1987) used USGS 30-m DEM's to produce slope, elevation, and aspect maps that were used as premaps for third-order soil

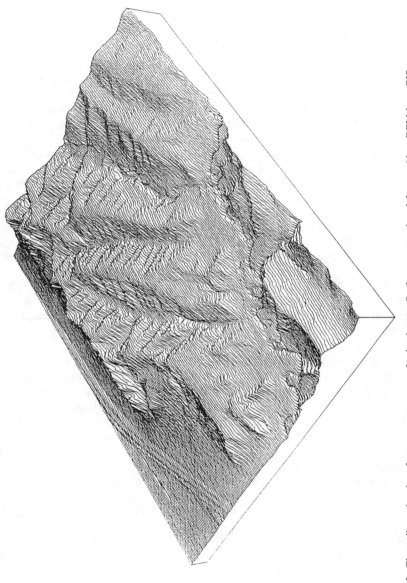

Fig. 15–2. Three-dimensional surface response curve of a landscape. Surface was produced from a 10-m DEM by a GIS.

surveys in Idaho, Nevada, and Wyoming. Favorable results were reported, but the authors concluded that slope and aspect maps generated from DEM's would have limited use in second-order soil surveys. They also reported that special-use maps from GIS had limited use in areas with slopes less than 5% and aspect less than 15°. Klingebiel et al. (1987) also concluded that: some topographic features that otherwise might have been overlooked on aerial photographs were identified in the premaps generated from DEM data; interpretations of premaps could be made during inclement weather; and premaps could be used by experienced field soil scientists to facilitate field mapping. Klingebiel et al. (1987) further concluded: DEM data could be used in conjunction with digitized soil surveys in resource databases, could be used to aid soil surveys in remote areas, and could be used in describing map unit composition and in locating field sampling sites. They urged that additional studies be conducted to investigate other applications of DEM data in soil surveys. As a result of this research, a guide for using DEM's in soil surveys was published (Klingebiel et al., 1988).

Slope classes of less than 4% were difficult to identify accurately with 30-m DEM (Klingebiel et al., 1987). Much of the terrain they investigated was steeply sloping with much relief. At first glance it would appear that DEM/GIS applications are limited in more gently sloping terrain with less relief. Alternatively, a finer-grid DEM database could be used. In Missouri, a 10-m DEM was used with GIS to accurately identify 79.4% of land in the 0 to 4% slope class in an undulating landscape with 19 m of vertical relief. For 0 to 4% category, 83% of the misclassified land was classified into a 5 to 9% slope class. These results indicate that DEM and GIS can be used in landscapes of more gentle relief to identify slope shapes and steepness. However, DEM with finer than 30-m resolution may be necessary in landscapes with gentle relief.

Use of Digital Elevation Models and Geographic Information Systems to Identify Landscape Attributes

A potentially important application of GIS and DEM is to identify land surface features that may be unnoticed in aerial photograph interpretation with a stereoscope, or which may be too small for separate delineation at the utilized map scale (Klingebiel et al., 1987). Possible uses for this information include development of premaps for mapping remote areas, production of slope/aspect maps as an aid in assessing forest site productivity, and development of land surface maps to increase precision of predicting crop yield in agricultural systems.

Determining map unit purity is another possible use. Frequency and characteristics of inclusions from random map units could be identified with transect analysis. Similar map units could be identified across the survey area. Patterns and characteristics of the measured inclusions could be extrapolated across the survey area with GIS/DEM. The transect and map unit data could be used together to predict map unit composition on a regional basis.

Figure 15-3 illustrates one way that GIS and DEM can be used together. The three-dimensional surface contour of a 16-ha field (Fig. 15-3A) was produced from a 10-m DEM. Microrelief within the field is distinctive in the graphic. The DEM data can also be used with GIS to produce a slope-class map (Fig. 15-3B) that shows more detail than could be displayed at a scale of 1:24 000 in a conventional soil survey field sheet. With GIS the slope-class map can be draped over the surface contour to show the relationships of microrelief to slopes classes (Fig. 15-3C). The contour intervals can be added to the figure and compared with slope classes. More sophisticated analyses are possible, but are beyond the scope of this chapter.

Such maps have great potential to reduce routine slope measurements by the soil surveyor. The same database that produced Fig. 15-3A and 15-3B could be used in calculations with the Universal Soil Loss Equation to model soil erosion. Land qualifying for the Conservation Reserve Program provisions of the Food Security Act of 1985 (House of Representatives Report 99-447) could be identified with GIS analysis of DEM data.

APPLICATION OF GEOGRAPHIC INFORMATION SYSTEMS TO EVALUATE SOIL PRODUCTIVITY

Forest Site Productivity Related to the Soil Series

The central concept of the soil map unit and the soil series is that soil pedons with similar chemical and physical properties or with similar management capabilities will be grouped together. Van Eck and Whiteside (1958) were among the first to investigate the utility of the soil series for forest-site productivity. They concluded that soil series "can be valuable tools in the prediction of site quality. . . . " However, their results contain neither statistical quantification of variability nor reports of site index variability within map units.

Other investigations have not been so positive. Farnsworth and Leaf (1965) studied sugar maple (*Acer saccharum* Marsh.) growth on four soil series in New York. They reported so much variation in tree growth within soil series that they recommended other methods of soil classification should be investigated. Van Lear and Hosner (1967) found "little, if any, usable correlation between soil mapping units and the site index of yellow-poplar (*Liriodendron tulipifera* L.) in southwestern Virginia." In southeastern Ohio, it was revealed that topographic features were more accurate predictors of black oak (*Quercus velutina* Lam.) site index than soil mapping units (Carmean, 1967). Site index of quaking aspen (*Populus tremuloides* Michx.) in Minnesota was reported to be "poorly related to soil mapping units" (Esu & Grigal, 1979). In Michigan, Shetron (1972) found "significant" site index differences within soil map units for jack pine, red oak, (*Quercus rubra* L.), and bigtooth aspen (*Populus grandidentata* Michx.), but not for sugar maple.

Jones (1969) wrote that site index variability within mapping units has been revealed to be so great in so many ecosystems that "soil series alone

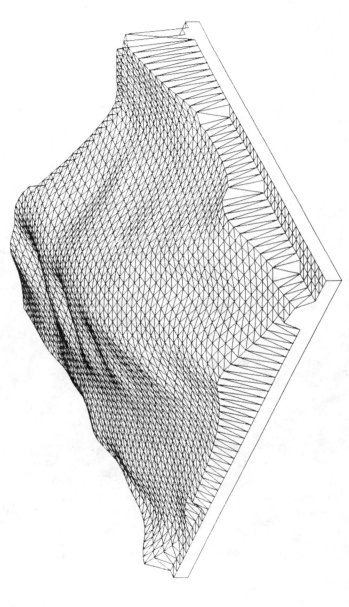

Fig. 15-3A. Three-dimensional surface contour of a 16-ha field in northwest Missouri. The surface contour was generated from 10-m DEM data.

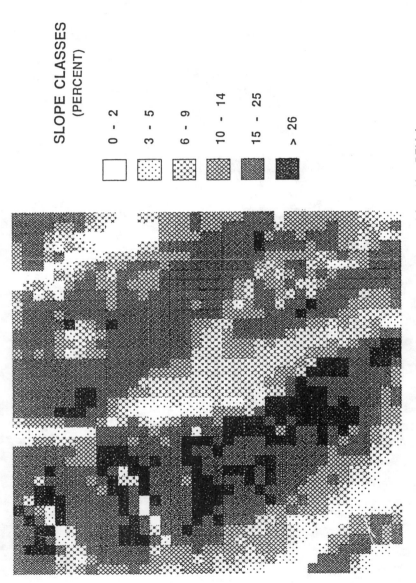

Fig. 15–3B. A slope-class map of the field depicted in Fig. 15–3A. Slope classes were calculated from 10-m DEM data.

GIS FOR SOIL SURVEY AND LAND USE

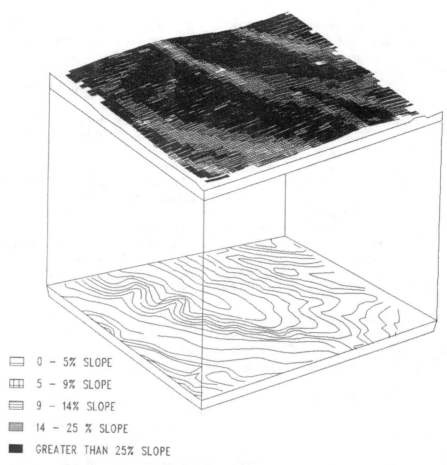

☐ 0 − 5% SLOPE
▦ 5 − 9% SLOPE
▤ 9 − 14% SLOPE
▩ 14 − 25 % SLOPE
■ GREATER THAN 25% SLOPE

Fig. 15–3C. The slope-class map draped over the three-dimensional surface contour of the field. Contour intervals are displayed below the slope-class map.

are too heterogeneous ecologically to serve as a basis for evaluating timber productivity. . . ." Jones offered the "more favorable" alternative of "landscape mapping" for forest productivity applications. Rowe (1984) has discussed the importance of basing forest site classification on the landform; and he noted the frequency with which moisture regime can be predicted from landform parameters. Daniels et al. (1971) and Ruhe (1975) have predicted that future attempts to sort soil units into more usable entities with reduced variability will rely more heavily upon geomorphology than have past efforts.

Soil/Site Factors Affecting White Oak (*Quercus alba* L.)

Growth of many forest tree species has been shown to be highly correlated to soil–site factors affecting potential evapotranspiration and the water-supplying capacity of the landscape. For purposes of illustration, we chose white oak because of its broad ecological amplitude, and because its site requirements overlap those of other commercially important forest species. Factors reported to affect growth of white oak include:

1. Thickness of the soil solum (Einspahr & McComb, 1951; McClurkin, 1963; Trimble & Weitzman, 1956; Yawney & Trimble, 1968).
2. Aspect (Carmean, 1965; Doolittle, 1957; Einspahr & McComb, 1951; Gaiser & Merz, 1951; Graney, 1977; Smalley, 1967; Trimble & Weitzman, 1956; Yawney, 1964; Yawney & Trimble, 1968).
3. Stand density (Gaiser & Merz, 1951).
4. Position on slope/distance from interfluve summit (Bowersox & Ward, 1972; Carmean, 1965; Della-Bianca & Olson, 1961; Doolittle, 1957; Einspahr & McComb, 1951; Gaiser & Merz, 1951; Graney, 1977; Hannah, 1968; Ike & Huppuch, 1968; McClurkin, 1963; Smalley, 1967; Trimble & Weitzman, 1956).
5. Thickness of the A horizon (Carmean, 1965; Doolittle, 1957; Gaiser & Merz, 1951; Hannah, 1968; Ike & Huppuch, 1968).
6. Soil texture (Bowersox & Ward, 1972; Graney, 1977; Hannah, 1968; Ike & Huppuch, 1968; McClurkin, 1963; Trimble & Weitzman, 1956).
7. Slope steepness (Bowersox & Ward, 1972; Carmean, 1965; Einspahr & McComb, 1951; Graney, 1977; Ike & Huppuch, 1968; Trimble & Weitzman, 1956; Yawney, 1964).
8. Slope shape (Graney, 1977; Hannah, 1968).
9. Elevation (Ike & Huppuch, 1968).
10. Slope length (Smalley, 1967).
11. Stoniness of the soil (Carmean, 1965).
12. Organic matter content of the A horizon (Della-Bianca & Olson, 1961).

Figure 15-4 displays growth of individual white oak and yellow-poplar stems on interfluve shoulders (10% slope) and north-facing slopes (35% slope) on the Cumberland Plateau in Tennessee (Hammer, 1986). Tree height at various ages was determined from stem analysis. Height growth of individual stems increased with distance downslope. Figure 15-5 reveals the polymorphic growth of yellow-poplar among geomorphic surfaces. The curves are mean growth by age for a minimum of four trees per site.

The obvious abilities to identify slope parameters and topographic factors important to tree growth with GIS/DEM technologies cannot be overlooked. Both site-specific or regional maps predicting potential tree growth as a function of soil properties and landscape position could be produced from DEM's using GIS. Remotely sensed infrared data might offer an additional data layer to assist in quantifying stand density parameters and soil moisture gradients. A potential site index could be produced with GIS in conjunction with DEM, soil survey, and knowledge of species site requirements. The conceptual blueprint is offered in Fig. 15-6.

GIS FOR SOIL SURVEY AND LAND USE

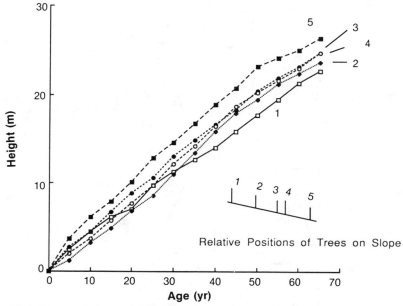

Fig. 15-4. Stem analysis curves of white oak growth on an interfluve summit.

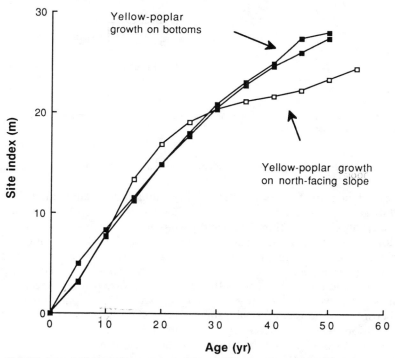

Fig. 15-5. Polymorphic growth of yellow-poplar on north-facing slopes and first-order bottoms. The curves are averages of growth of four trees per landform.

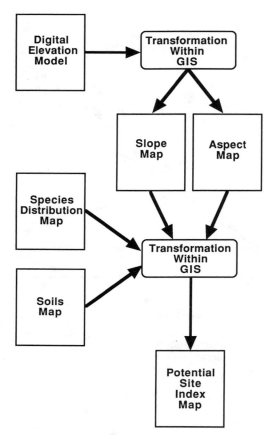

Fig. 15-6. Conceptual framework for generating a potential site index map with a GIS and several data sources.

Assessing Potential Agricultural Crop Yield as a Function of Landscape

Soil–landscape factors affecting plant–available soil water can influence yield of agricultural crops. Only recently has attention been focused on this topic. Traditionally, within-field yield variability has been attributed to soil erosion rather than to naturally occurring variability in soil and water distribution related to geomorphic surface shape, stratigraphy, or soil-profile attributes (Daniels et al., 1987). The effects of landscape position on agricultural crop yield have not received the attention accorded forest species, but Daniels and coworkers have investigated yield variability for annual agricultural crops. Simmons et al. (1989) showed corn yield response to slope shape and position in Atlantic Coastal Plain soils. Daniels et al. (1989) reported crop yield variability unrelated to soil erosion class in Piedmont soils. In Missouri, Boudeman (1989) reported temporal and spatial variability in forage legume yield related to slope shape and position. Daniels and Nelson (1987) have suggested that multidisciplinary research that carefully considers stratigraphy, geomorphology, and hydrology is necessary if causal relationships affecting crop growth in natural landscapes are to be disentangled.

More attention should be focused on soil-landscape-related variability in agronomic crop yield. Meaningful results of such research will require multidisciplinary research over several years. Crop yield, management practices, climatic variability, soil properties, geomorphic surface attributes, stratigraphy, and remotely sensed crop and soils data should be integrated with GIS in long-term research projects.

APPLICATIONS OF GEOGRAPHIC INFORMATION SYSTEMS FOR SOIL-LANDSCAPE RESEARCH

Cline (1961) noted that a scientific discipline is reflected in the conceptual model that composes the mental image of the entity under study. Ideally, the model of a discipline is tested by scientific experiments and is revised to reflect increasing knowledge as quantitative data are acquired through research. One might cautiously argue that new or greatly revised general models of soil genesis have not been manifested of late.

Simonson's (1959) "Generalized Model of Soil Genesis" provided for process-related pedology research a conceptual framework for investigating and defining soil properties used as diagnostic criteria for the *Soil Taxonomy* (Soil Survey Staff, 1975). Simonson's model also served as a template for defining the genetic pathways used to properly link morphologically related soils (Cline & Johnson, 1963). Johnson (1963) defined the soil individual to be examined and classified from the continuum of soils on the earth's mantle. Arnold (1965) introduced to soil scientists the geologists' theory of multiple-working hypotheses, an important attribute for the conceptual approach to morphologically similar soils produced from different environments. Cline (1961) observed that a result of the focus upon *Soil Taxonomy* was to have changed the conceptual model of soil " . . . from one in which the whole is emphasized and its parts are loosely defined and indistinct to one in which the parts are sharply in focus and the whole is an organized collection of parts." Rowe (1984) noted that this emphasis on taxonomy caused pedologists to think of soils as "things unto themselves" and lamented the loss of perspective accompanying this focus.

The conceptual framework of pedologists shifted predominantly from the continuum concept (Marbut, 1935) to the soil individual, although the soil landscape continued to receive attention. Jenny's (1941) factors of soil formation remained important as a conceptual framework for soil genesis. Field research was often "stratified" by attempting to hold constant one or more of the soil-forming factors (Dickson & Crocker, 1952). Other investigators criticized Jenny's model, citing the inability to mathematically solve its integral relationships (Runge, 1973; Huggett, 1975).

The catena concept (Milne, 1936) received attention. The climatoposequence became a conceptual framework for research. For example, both depth to fragipan and soil drainage were related to topographic position on a drumlin in New England (Veneman & Bodine, 1982). Soil series and soil drainage patterns were related to topographic position in a loess-mantled glacial till landscape in Indiana (Evans & Franzmeier, 1986). Soil N and other soil properties varied systematically with land-surface shape and topographic position in an Iowa native prairie (Aandahl, 1948).

Growth and distribution of vegetation often is associated with soil and landscape features. Both soil properties and ground flora communities were related to landscape position in a toposequence in upper Michigan (Pregitzer et al., 1983). On the Allegheny Plateau in Ohio, northeast-facing slopes had a more mesophytic vegetation, thicker A horizons, higher pH, lower C/N ratios, and less well-expressed E horizons than soil on southwest-facing slopes (Finney et al., 1962).

Geological strata often influence landform evolution and consequently influence distribution of soil water and vegetation. Soil series, soil drainage, and patterns of native vegetation were correlated to geomorphic surfaces related to bedrock stratigraphy and to colluvial periglacial landforms in the Allegheny Plateau in Pennsylvania (Aguilar & Arnold, 1985). Huddleston and Riecken (1973) studied soil chemical and physical properties in dissected deep loess deposits in Iowa. They concluded that the landscape was a record of interactions between "pedological redistribution" of P, clay and Fe, and the geological process of surface modification by landscape-forming processes.

It is interesting to note that in a fraction of these studies the distributional patterns of native vegetation were observed in their relationships to soil and landscape parameters controlling plant-available soil–water and potential evapotranspiration stress. Field ecologists have long noted the frequent correlation between parent materials, landform parameters affecting soil-water distribution, soil properties, and patterns of vegetation distribution and production (Ayyad & Dix, 1964; Hopkins, 1951; Weaver & Albertson, 1956; Whittaker, 1967). Have many pedologists forgotten the relationships between soils and plant communities? Daniels and Nelson (1987) urged that

> We must abandon the idea that soils are independent entities occurring at specific points and consider that each part of a landscape is related. Each part is affected by and affects the adjacent parts, especially those downslope or in the direction of the surface hydraulic gradient. What is needed is a better understanding of the soil–plant environment, and this must include the physical basis for soil variability . . . we cannot predict impacts until we understand processes responsible for soil variability on landscapes and processes responsible for differential crop yields on those landscapes.

Even as attention was focusing upon the soil individual, Knox (1965) expressed the ". . . need for a concept of soil corresponding to some kind of soil landscape unit." Cline (1961) predicted that ". . . the impact (of geomorphology) will . . . be a major force in the development of models of the future." Ruhe and his coworkers introduced the importance of climatic variability, paleosols, and stratigraphy in soil–landscape analysis (Ruhe, 1956; Ruhe & Scholtes, 1956; Ruhe et al., 1967). Ruhe subsequently expanded the scope of his research to include open (Ruhe & Walker, 1968) and closed (Walker & Ruhe, 1968) watersheds, thus welding the soil landscape to hydrological processes.

Dan and Yaalon (1968), recognizing the frequent correlation between soils and geomorphic units within a landscape, coined the term "pedomorphic surface" for a landscape in which soil and relief are related genetically and have coevolved. They noted that in such landscapes a specific "pedomorphic form" (horizon sequence) accompanies the individual landforms within the landscape. Lanyon and Hall (1983) described the lack of attention given to morphology of land surfaces and developed an algorithm to describe landforms. They noted that the accuracy of calculated land surfaces was a function of the frequency of data points and the validity of the assumption of linearity in parameters between elevation points. The types of measurements they made are now possible with commercially available GIS software. Arnold (1987) described soil landscapes as "significant hydrologic and pedogenic entities" and said that studies emphasizing soil properties with landscape positions that could model and predict interactions within catenary units would "change guidelines for the soil survey."

Yaalon (1975) discussed the preponderance of two-dimensional catenary studies and lamented the paucity of research investigating soil landscapes. He expressed the hope that "As computer methods enable easier curve fittings, and as better and more quantitative data on the nature and development of slopes are generated this will also . . . be reflected in soil landscape systems studies . . . with eventually some generally valid rules . . ." Huggett (1975) stated that the soil system, with the variety of temporal components, is too complex to be described with isomorphic models, and he recommended that pedologists adapt Odum's (1971) "macroscopic" viewpoint, focusing upon the system as a whole. Rowe (1984) chastised pedologists for forgetting that neither the relationships of soils nor plant communities with the environment can be understood alone. Both soils and vegetation are welded by their relationships to geomorphic surfaces that determine the flux of energy and water through both soils and plants. He said ". . . the problem—and I think there is one—will be solved when pedologists become ecologists again, perceiving their soils as the rooting medium and detritus component of ecosystems, shifting their focus to the three-dimensional characteristics of soils and their environmental significance, particularly with respect to moisture regime."

The message is clear. The need is great for multidisciplinary research that focuses upon the multidimensional (in both space and time) nature of the soil–plant interface. Research must address the processes responsible for soil spatial and temporal variability and must quantify the effects of this variability on human soil management practices. The variability of mapped soil series must be more precisely quantified with spatial and temporal data statistics as a first step (Baker & Rogowski, 1987). Predictive models must be developed and tested under field conditions if mankind hopes to properly manage the soil resource into the future. Cline (1961) observed that "new and more precise knowledge . . ." is often acquired with "new instruments and techniques." The unique attributes of GIS, DEM, and remote sensing seem well-suited to the challenge.

CONCLUSIONS

Possible Uses of Geographic Information Systems/Digital Elevation Models as Tools in Soil and Crop Research

Applications of GIS/DEM appear limited only by the investigators' imaginations. However, a list of potential uses of these technologies is presented for consideration. In listing these potential applications, we have not attempted to separate pedological uses from soil survey applications. The two are, or should be, closely intertwined, particularly where landscape analysis is considered.

Geographic Information Systems as an Organizational Framework for Landscape Analysis

The GIS/DEM technologies are suitable for incorporating data in a framework for analysis of entire watersheds or landscapes. Selection of the proper DEM/DTM resolution will be important, but should allow precise quantification and analysis of geomorphic surfaces that otherwise might not be considered. Geomorphic surfaces could be identified, described, and compared with the spatial analysis capabilities of GIS. The scope of such studies could range from single watersheds to regional drainage basins.

Important components such as stratigraphy, geology, vegetation, management practices, soil properties, climate attributes (solar radiation, temperature/degree days, seasonal precipitation, etc.) seasonal soil–water distribution, and geomorphology could be included within landscape-analysis databases. Special-use maps could be produced to display the spatial relationships of these layers of the ecosystem matrix.

Predictive Models for Crop Yield and Soil Genesis

From the data described above, predictive models of soil processes and crop yields could be developed. Comparable areas for field testing the models could be identified with the aid of GIS/DEM. Heterogeneity in crop yield across a landscape could be analyzed with this system, and long-term yield variability could be correlated with landscape and soil patterns.

Root-growth response could be compared to seasonal soil–water distribution and landform-related soil properties. The importance of these data for development of a soil-based productivity index should not be overlooked (Henderson et al., 1990).

Seasonal distribution of soil water and soil solution composition could be compared with stratigraphic, geomorphic, and pedologic variables to more precisely quantify the relationships of soil temporal properties to the three-dimensional patterns of soil–landscape features. Nutrient distribution in the landscape could be compared with patterns of temporal soil–water availability, and could further be compared to patterns of organic matter accumulation and decay.

Spatial Analysis of Soils and Geomorphic Surfaces

The size, purity, and distribution of map units could more readily be compared to areal distribution of stratigraphy, geomorphic surfaces, geology, and vegetation. Soil variability across geomorphic surfaces could be precisely analyzed for relationships to surface shape and slope position. Patterns of limiting and nonlimiting soil inclusions could be spatially compared to landform surface-shape patterns and stratigraphy. Patterns of soil–water movement in watersheds could be related to pedologic features and geomorphology.

Information acquired through the comparisons mentioned above could then be extrapolated across regional areas using GIS and DEM. Representative areas for ground-truth measurements then could be identified with the aid of GIS.

Long-term Data Acquisition, Storage, and Analysis

The ability to repeatedly add information to existing databases and to constantly compare the patterns of temporal factors with the more enduring attributes of surface shape and stratigraphy are most appealing. Annual crop yields in a watershed or the mean annual increment of forest growth, for example, could be accumulated for years, then spatially analyzed.

Annual climatic variables could also be retained for comparison with crop yield fluctuations, temporal nutrient flux, and seasonal distribution of soil water in the landscape.

A Call for Action

Soil–landscape relationships generally have been studied as transects or in the two-dimensional context of the catena. Soils frequently have been studied as entities unto themselves; and relationships of soils and landforms to vegetation have not received the attention they deserve. Increasingly, demands are being voiced for more precise quantification of soil survey map units and for pedological models addressing the soil–plant interface and soil–landscape relationships. Such studies should include temporal and spatial soil variability, more precise quantification of relationships between landforms and soils, and consideration of patterns and pathways of soil–water distribution.

A science evolves only as new quantitative data are obtained and incorporated. The technologies of GIS and DEM are proposed as analytical tools capable of integrating spatial and temporal data, conducting spatial analyses of layered databases, and producing special-use maps that constitute new information composed of combinations of the various data. The best potential use of GIS is as a powerful aid to increase our understanding of the resources being studied. As such, GIS and DEM offer new alternatives for conducting field research. They are methods to utilize existing knowledge and to store, monitor, and analyze data from multidisciplinary, long-term research. We encourage our colleagues to investigate the potential of GIS for soil survey, field pedology and geomorphological research. The potential exists to bring a higher level of precision to quantitative description and prediction of the spatial attributes of soils, sediments, and landforms.

APPENDIX A

Sources of Information for Geographic Information System Applications and Technology

E.G. Pecora Symposium Proceedings; published annually from 1975 to present by the American Society for Photogrammetry and Remote Sensing, 210 Little Falls St., Falls Church, VA 22046.

Proceedings of Annual International Conference, Exhibits, and Workshops on Geographic Information Systems (GIS/LIS 1986, 1987, 1988); published by the American Society for Photogrammetry and Remote Sensing and the American Congress on Surveying and Mapping. American Society for Photogrammetry and Remote Sensing, 210 Little Falls St., Falls Church, VA 22046.

GIS Technical Papers, an annual publication from the American Society for Photogrammetry and Remote Sensing and the American Congress on Surveying and Mapping. American Society for Photogrammetry and Remote Sensing, 210 Little Falls St., Falls Church, VA 22046.

International Geographic Information System Symposium: The Research Agenda. Three volumes, edited by Robert Aangeenbrug and Yale Schiffman; and published by the Association of American Geographers, 1710 16th St., Washington, DC.

Proceedings 1989 National Conference of Geographic Information Systems; published by the Canadian Institute of Surveying and Mapping, Box 5378, Station F, Ottawa, Canada K2C 3J1.

Proceedings of the Annual Conference of Urban and Regional Information Systems Association; published by URISA, 319 C. St. S.E., Washington, DC 20003.

REFERENCES

Aandahl, A.R. 1948. The characterization of slope positions and their influence on the total nitrogen content of a few virgin soils of western Iowa. Soil Sci. Soc. Am. Proc. 13:449–454.

Aguilar, R., and R.W. Arnold. 1985. Soil-landscape relationships of a climax forest in the Allegheny High Plateau, Pennsylvania. Soil Sci. Soc. Am. J. 49:695–701.

Ahern, F.J., W.J. Bennett, and E.G. Kettela. 1985. Surveying spruce budworm defoliation with an airborne pushbroom scanner: A summary. p. 228-235. *In* Precora 10. Remote sensing in forest and range resource management. Proc. Am. Soc. Photogramm. Remote Sens., Fort Collins, CO. 20-22 Aug. 1985. Am. Soc. Photogramm. Remote Sens., Falls Church, VA.

Arnold, R.W. 1965. Multiple working hypotheses in soil genesis. Soil Sci. Soc. Am. Proc. 29:717-724.

Arbikdm R,W, 1987. The future of the soil survey. p. 261-268. *In* L.L. Boersma et al. (ed.) Future development in soil science research. SSSA, Madison, WI.

Ayyad, M.A.G., and R.L. Dix. 1964. An analysis of a vegetation-microenvironmental complex on prairie slopes in Sasketchewan. Ecol. Monogr. 34:421-441.

Baker, D.E., and A.S. Rogowski. 1987. Database requirements for expert systems in land resource management. p. 115-124. *In* L.L. Boersma et al. (ed.) Future developments in soil science research. SSSA, Madison, WI.

Balser, R., and K. Fila. 1987. Planning for terrain resource information management (TRIM) in British Columbia. p. 91-100. *In* GIS '87—San Francisco Vol. I. Proc. 2nd Ann. Int. Conf. on GIS., San Francisco, CA. 26-30 Oct. 1987. Am. Soc. Photogramm. Remote Sens., Falls Church, VA.

Berry, J. 1986. Using a microcomputer system to spatially characterize effective timber accessibility. p. 273-283. *In* Proc. Geographic information systems workshop, Atlanta, GA. 1-4 Apr. 1986. Am. Soc. Photogramm. Remote Sens., Falls Church, VA.

Best, R.G., and F.C. Westin. 1984. GIS for soils and rangeland management. p. 70-74. *In* Precora 9. Spatial information technologies for remote sensing today and tomorrow. Proc. Am. Soc. Photogramm. Remote Sens., Sioux Falls, SD. 2-4 Oct. 1984. Sioux Falls, SD. Am. Soc. Photogramm. Remote Sens., Falls Church, VA.

Boudeman, J.W. 1989. Above- and below-ground production of three forage legumes related to soil properties and a soil-based productivity index. M.S. thesis. University of Missouri-Columbia.

Bowersox, T.W., and W.W. Ward. 1972. Prediction of oak site index in the Ridge and Valley region of Pennsylvania. For. Sci. 18:192-195.

Burrough, P.A. 1986. Principles of geographical information systems for land resources assessment. Clarendon Press, Oxford.

Carmean, W.H. 1965. Black oak site quality in relation to soil and topography in southeastern Ohio. Soil Sci. Soc. Am. Proc. 29:308-312.

Carmean, W.H. 1967. Soil survey refinements for predicting black oak site quality in southeastern Ohio. Soil Sci. Soc. Am. Proc. 31:805-810.

Cline, M.G. 1961. The changing model of soil. Soil Sci. Soc. Am. Proc. 25:442-446.

Cline, A.J., and D.D. Johnson. 1963. Threads of genesis in the seventh approximation. Soil Sci. Soc. Am. Proc. 27:220-222.

Cowen, D.J. 1988. GIS vs. CAD vs. DBMS: What are the differences. Photogramm. Eng. Remote Sens. 54:1551-1556.

Dan, J., and D.H. Yaalon. 1968. Pedomorphic forms and pedomorphic surfaces. p. 577-584. *In* J.W. Holmes et al. (ed.) Int. Congr. Soil Sci. Trans. 9th, Vol. 4. Adelaide, Australia. Elsevier Publ. Co., New York.

Daniels, R.B., E.E. Gamble, and J.G. Cady. 1971. The relation between geomorphology and soil morphology and genesis. Adv. Agron. 23:51-88.

Daniels, R.B., J.W. Gilliam, D.K. Cassel, and L.A. Nelson. 1987. Quantifying the effects of past soil erosion on present soil productivity. J. Soil Water Conserv. 42:183-187.

Daniels, R.B., J.W. Gilliam, D.K. Cassel, and L.A. Nelson. 1989. Soil erosion has limited effect on field scale crop productivity in the Southern Piedmont. Soil Sci. Soc. Am. J. 53:917-920.

Daniels, R.B., and L.A. Nelson. 1987. Soil variability and productivity: Future developments. p. 279-291. *In* L.L. Boersma et al. (ed.) Future developments in soil science research. SSSA, Madison, WI.

Della-Bianca, L., and D.F. Olson, Jr. 1961. Soil-site studies in Piedmont hardwood and pine-hardwood upland forests. For. Sci. 7:320-329.

Dickson, B.A., and R.L. Crocker. 1952. A chronosequence of soils and vegetation near Mt. Shasta, California I. Definition of the ecosystem investigated and features of the plant succession. J. Soil Sci. 4:123-141.

Doolittle, W.T. 1957. Site index of scarlet and black oak in relation to southern Appalachian soil and topography. Forest Sci. 3:114-124.

Einspahr, D., and A.L. McComb. 1951. Site index of oaks in relation to soil and topography in northeastern Iowa. J. For. 49:719-723.

Elassal, A.A., and V.M. Caruso. 1984. Digital elevation models. U.S. Geol. Surv. Circ. 895-B.

Esu, I.E., and D.F. Grigal. 1979. Productivity of quaking aspen. Soil Sci. Soc. Am. J. 43:1189-1192.

Evans, C.V., and D.P. Franzmeier. 1986. Saturation, aeration, and color patterns in a toposequence of soils in north-central Indiana. Soil Sci. Soc. Am. J. 50:975-980.

Farnsworth, C.E., and A.L. Leaf. 1965. An approach to soil-site problems: Sugar maple-soil relations in New York. In C.T. Youngberg (ed.) Forest soil relationships in North America. Proc. North American For. Soils Conf., 2nd. Oregon State Univ. Press, Corvallis, OR.

Finney, H.R., N. Holowaychuk, and M.R. Heddleson. 1962. The influence of microclimate on the morphology of certain soils of the Allegheny Plateau of Ohio. Soil Sci. Soc. Am. Proc. 26:287-292.

Fitzgerald, C. 1987. Justifying a GIS for a local planning agency. p. 10-18. In GIS '87—San Francisco. Proc. 2nd Ann. Int. Conf. GIS., San Francisco, CA. 26-30 Oct. 1987. Am. Soc. Photogramm. Remote Sens., Falls Church, VA.

Gaiser, R.N., and G.W. Merz. 1951. Stand density as a factor in estimating white oak site index. J. For. 49:572-574.

Graney, D.L. 1977. Site index prediction for red oaks and white oaks in Boston Mountains of Arkansas. USDA Forest Serv. Res. pap. SO-139.

Hammer, R.D. 1986. Soil variability, soil water, and forest tree growth on three forested Cumberland Plateau landtypes. Ph.D. diss. Univ. Tennessee, Knoxville (Diss. Abstr. 8708772).

Hannah, P.R. 1968. Estimating site index for white and black oaks in Indiana from soil and topographical factors. J. For. 66:412-417.

Henderson, G.S., R.D. Hammer, and D.F. Grigal. 1990. Can measurable soil properties be integrated into a framework for characterizing forest productivity? p. 137-154. In S.P. Gessel et al. (ed.) Sustained productivity of forest land. Proc. 7th North Am. For. Soils Conf., Vancouver, BC. 24-28 1988. Univ. British Columbia, Vancouver, Canada.

Hermann, K., and B. Floyd. 1986. Development of a state-wide digitized soil resource inventory. p. 368-377. In Proc. Geographic Information Systems Workshop, Atlanta, GA. 1-4 Apr. 1986. Am. Soc. Photogramm. Remote Sens., Falls Church, VA.

Hopkins, H.H. 1951. Ecology of the native vegetation of the loess hills in central Nebraska. Ecol. Monogr. 21:125-147.

Horvath, E.H., D.F. Post, and J.B. Kelsey. 1984. The relationships of Landsat digital data to the properties of Arizona rangelands. Soil Sci. Soc. Am. J. 48:1331-1334.

Huddleston, J.H., and F.F. Riecken. 1973. Local soil-landscape relationships in western Iowa: I. Distributions of selected chemical and physical properties. Soil Sci. Soc. Am. Proc. 37:264-270.

Huggett, R.J. 1975. Soil landscape systems: A model of soil genesis. Geoderma 13:1-22.

Ike, A.F., and C.D. Huppuch. 1968. Predicting tree height growth from soil and topographic site factors in the Georgia Blue Ridge Mountains. Georgia For. Res. Council Res. Pap. 54.

Jenny, H. 1941. Factors of soil formation—a system of quantitative pedology. McGraw-Hill, New York.

Johnson, W.M. 1963. The pedon and the polypedon. Soil Sci. Soc. Am. Proc. 27:212-215.

Jones, J.R. 1969. Review and comparison of site evaluation methods. USDA For. Serv. Res. Pap. RM-51.

Klingebiel, A.A., E.H. Horvath, D.G. Moore, and W.U. Reybold. 1987. Use of slope, aspect, and elevation maps derived from digital elevation model data in making soil surveys. p. 77-90. In W.U. Reybold and G.W. Peterson (ed.) Soil survey techniques. SSSA Spec. Publ. 20. SSSA, Madison, WI.

Klingebiel, A.A., E.H. Horvath, W.U. Reybold, D.G. Moore, E.A. Fosnight, and T.R. Loveland. 1988. A guide for the use of digital elevation model data for making soil surveys. USDA-SCS and U.S. Dep. of Interior, U.S. Geol. Surv. Open-File Rep. no. 88-102.

Knox, E.G. 1965. Soil individuals and soil classification. Soil Sci. Soc. Am. Proc. 29:79-84.

Lanyon, L.E., and G.F. Hall. 1983. Land-surface morphology: 1. Evaluation of a small drainage basin in eastern Ohio. Soil Sci. 136:291-299.

Lee, K., G.B. Lee, and E.J. Tyler. 1988. Thematic mapper and digital elevation modeling of soil characteristics in hilly terrain. Soil Sci. Soc. Am. J. 52:1104-1107.

Logan, T.J., D.R. Urgan, J.R. Adams, and S.M. Yaksich. 1982. Erosion control potential with conservation tillage in the Lake Erie Basin: Estimates using the Universal Soil Loss Equation and the Land Resource Information System (LRIS). J. Soil Water Conserv. 37:50-55.

Marbut, C.F. 1935. Soils of the United States. *In* USDA atlas of American agriculture.

McClurkin, D.C. 1963. Soil-site index predictions for white oak in north Mississippi and west Tennessee. For. Sci. 9:108-113.

Mead, R.A., and E.B. Rowland. 1987. Selecting sand pine stands for harvest using a GIS. p. 705. *In* GIS '87—San Francisco Vol. I. Proc. 2nd Ann. Int. Conf. GIS., San Francisco, CA. 26-30 Oct. 1987. Am. Soc. Photogramm. Remote Sens., Falls Church, VA.

Milne, G. 1936. A provisional soil map of East Africa. Amani Memoirs, East Afric. Agric. Res. Stn., Tanganyika Territory.

Murray, R.J., and D.A. Leckenby. 1985. Elk habitat evaluation using distance-mapped Landsat data. p. 346-355. *In* Precora 10. Remote sensing in forest and range resource management. Proc. Am. Soc. Photogram. Remote Sens., Fort Collins, CO. 20-22 Aug. 1985. Am. Soc. Photogramm. Remote Sens., Falls Church, VA.

Odum, H.T. 1971. Environment, power, and society. John Wiley & Sons, Inc., New York.

Parker, H.D. 1987. What is a geographic information system? p. 72-80. *In* GIS '87—San Francisco Vol. I. Proc. 2nd Ann. Int. Conf. GIS., San Francisco, CA. 26-30 Oct. 1987. Am. Soc. Photogramm. Remote Sens., Falls Church, VA.

Parks, B.O., G.A. Simmons, and S. Gage. 1987. A spatial model and geographic information system for statewide assessment of risks due to gypsy moth infestation in Michigan. p. 707. *In* GIS '87—San Francisco Vol. I. Proc. 2nd Ann. Int. Conf. GIS., San Francisco, CA. 26-30 Oct. 1987. Am. Soc. Photogramm. Remote Sens., Falls Church, VA.

Pregitzer, K.S., B.V. Barnes, and G.D. Lemme. 1983. Relationship of topography to soils and vegetation in an upper Michigan ecosystem. Soil Sci. Soc. Am. J. 47:117-123.

Rhind, D. 1977. Computer-aided cartography. Trans. Int. Br. Geogrs. 2:71-96.

Rowe, J.S. 1984. Forestland classification: Limitations of the use of vegetation. p. 132-147. *In* J.G. Bockheim (ed.) Forest land classification: Experiences, problems, perspectives. Proc. Symp., Univ. Wisconsin, Madison. 18-20 Mar. 1984.

Ruhe, R.V. 1956. Geomorphic surfaces and the nature of soils. Soil Sci. 82:441-455.

Ruhe, R.V. 1975. Geomorphology. Houghton Mifflin Co., Boston, MA.

Ruhe, R.V., R.B. Daniels, and J.G. Cady. 1967. Landscape evolution and soil formation in southwestern Iowa. USDA Tech. Bull. 1349.

Ruhe, R.V., and W.H. Scholtes. 1956. Age and development of soil landscapes in relation to climatic and vegetational changes in Iowa. Soil Sci. Soc. Am. Proc. 20:264-273.

Ruhe, R.V., and P.H. Walker. 1968. Hillslope models and soil formation: I. Open systems. p. 551-560. *In* J.D. Hobes et al. (ed.) Trans. Int. Congr. Soil Sci., 9th, Vol. 4, Adelaide, Australia.

Runge, E.C.A. 1973. Soil development sequence and energy models. Soil Sci. 115:183-193.

Rush, W.R., S.M. Howard, and W.D. Harrison. 1985. Mapping rangeland vegetation using Landsat MSS digital data for resource management planning. p. 102-108. *In* Precora 10. Remote sensing in forest and range resource management. Proc. Am. Soc. Photogramm. Remote Sens., Fort Collins, CO. 20-22 Aug. 1985. Am. Soc. Photogramm. Remote Sens., Falls Church, VA.

Shetron, S.G. 1972. Forest site productivity among soil taxonomic units in northern lower Michigan. Soil Sci. Soc. Am. Proc. 36:358-363.

Simmons, F.W., D.K. Cassel, and R.B. Daniels. 1989. Landscape and soil property effects on corn grain yield response to tillage. Soil Sci. Soc. Am. J. 53:534-539.

Simonson, R.W. 1959. Outline of a generalized theory of soil genesis. Soil Sci. Soc. Am. Proc. 23:152-156.

Smalley, G.W. 1967. Soil-site relations of upland oaks in north Alabama. USDA For. Serv. Res. Note SO-64.

Smith, S.M., H. Schreier, and R. Wiart. 1987. Agricultural field management with microcomputer-based GIS and image analysis systems. p. 585-594. *In* GIS '87—San Francisco Vol. I. Proc. 2nd Ann. Int. Conf. GIS., San Francisco, CA. 26-30 Oct. 1987. Am. Soc. Photogramm. Remote Sens., Falls Church, VA.

Soil Survey Staff. 1975. Soil taxonomy: A basic system of soil classification for making and interpreting soil surveys. Agric. Handb. 436. U.S. Gov. Print. Office, Washington, DC.

Stoner, E.R., and M.F. Baumgardner. 1981. Characteristic variations in reflectance of surface soils. Soil Sci. Soc. Am. J. 45:1161-1165.

Su, H., M.D. Ransom, and E.T. Kanemasu. 1989. Detecting soil information on a native prairie using Landsat TM and SPOT satellite data. Soil Sci. Soc. Am. J. 53:1479–1483.

Swann, R., A. Westwell-Roper, S. Wood, R. Rose, and W. Laing. 1987. The automated extraction of digital terrain models from satellite imagery. *In* Proc. 13th Ann. Conf., Int. Soc. Prog. Remote Sens., Nottingham, England. 7–11 Sept. 1987.

Swann, R., D. Kauffman, and B. Sharpe. 1988. Results of automated digital elevation model generation from SPOT satellite data. *In* Proc. 14th Ann. Conf., Int. Soc. Prog. Rem. Sens., Kyoto, Japan. 1–10 July, 1988.

Thompson, D.R., and K.E. Henderson. 1984. Detecting soils under cultural vegetation using digital Landsat thematic mapper data. Soil Sci. Soc. Am. J. 48:1316–1319.

Tomlinson, R.F. (ed.) 1972. Geographical data handling. IGU Comm. Geogr. Data Sens. and Process., Ottowa, Canada.

Trimble, G.R., and S. Weitzman. 1956. Site index studies of upland oaks in the northern Appalachians. For. Sci. 2:162–173.

Van Eck, W.A., and E.P. Whiteside. 1958. Soil classification as a tool in predicting forest growth. p. 218–226. *In* D.P. White (ed.) Proc. 1st N. Am. For. Soils Conf., Mich. State Univ., East Lansing, MI.

Van Lear, D.H., and J.F. Hosner. 1967. Correlation of site index and soil mapping units poor for yellow-poplar in southwest Virginia. J. For. 65:22–24.

Veneman, P.L.M., and S.M. Bodine. 1982. Chemical and morphological soil characteristics in a New England drainage-toposequence. Soil Sci. Soc. Am. J. 46:359–363.

Walker, P.H., and R.V. Ruhe. 1968. Hillslope models and soil formation. II. Closed systems. p. 561–568. *In* J.W. Holmes et al. (ed.) Trans. Int. Congr. Soil Sci., 9th, Vol. 4, Adelaide, Australia.

Walsh, S.J. 1985. Geographic information systems for natural resources management. J. Soil Water Conserv. 40:202–205.

Weaver, J.E., and F.W. Albertson. 1956. Grasslands of the great Plains. Johnsen Publ. Co., Lincoln, NE.

Whitehead, C. 1986. Tennessee aquatic database system. p. 128–132. *In* Proc. Geographic Inform. Systems Workshop, Atlanta, GA. 1–4 Apr. 1986. Am. Soc. Photogramm. Remote Sens., Falls Church, VA.

Whittaker, R.H. 1967. Gradient analysis of vegetation. Biol. Rev. 42:207–264.

Yaalon, D.H. 1975. Conceptual models in pedogenesis: Can soil-forming functions be solved? Geoderma 14:189–205.

Yawney, H.W. 1964. Oak site index on Belmont limestone soils in the Allegheny Mountains of West Virginia. USDA For. Serv. Res. Pap. NE-30.

Yawney, H.W., and G.R. Trimble, Jr. 1968. Oak soil-site relationships in the Ridge and Valley region of West Virginia and Maryland. USDA For. Serv. Res. Pap. NE-96.

Young, F.J., J.M. Maatta, and R.D. Hammer. 1991. Confidence intervals for soil properties within map units. p. 213–229. *In* M.J. Mausbach and L.P. Wilding (ed.) Spatial variabilities of soils and landforms. SSSA Special Publ. 28. SSSA, Madison, WI.

Zakrjewska, B. 1967. Trends and methods in land form geography. Ann. Assoc. Am. Geogr. 57:128–165.